Alternative Development Strategies and Appropriate Technology
(Pergamon Policy Studies—30)

Pergamon Policy Studies on the New International Economic Order

Haq   *A New Strategy For North—South Negotiations*

Meagher   *An International Redistribution of Wealth and Power: A Study of the Charter of Economic Rights and Duties of States*

Menon   *Bridges Across the South: Technical Cooperation Among Developing Countries*

UNITAR/CEESTEM Library on NIEO

Laszlo et al   *The Objectives of the New International Economic Order*

Laszlo/Kurtzman   *Europe and the New International Economic Order*

Laszlo/Kurtzman   *Political and Institutional Issues of the New International Economic Order*

Laszlo/Kurtzman   *The Structure of the World Economy and the New International Economic Order*

Laszlo/Kurtzman   *The United States, Canada and the New International Economic Order*

Laszlo/Kurtzman/Tikhomirov   *The Soviet Union, Eastern Europe and the New International Economic Order*

Laszlo/Lozoya et al   *The Implementation of the New International Economic Order*

Laszlo/Lozoya et al   *The Obstacles Confronting the New International Economic Order*

Laszlo/Lozoya et al   *World Leadership and the New International Economic Order*

Lozoya et al   *Africa, Middle East and the New International Economic Order*

Lozoya et al   *Asia and the New International Economic Order*

Lozoya et al   *International Trade, Industrialization and the New International Economic Order*

Lozoya et al   *Latin America and the New International Economic Order*

Lozoya et al   *The Social and Cultural Issues of the New International Economic Order*

Lozoya/Bhattacharya   *The Financial Issues of the New International Economic Order*

Lozoya/Estevez/Green et al   *Alternative Views of the New International Economic Order: A Survey and Analysis of Major Academic Research Reports*

ON THE NEW INTERNATIONAL
ECONOMIC ORDER

Alternative Development Strategies and Appropriate Technology

Science Policy for an Equitable World Order

Romesh K. Diwan
Dennis Livingston

Pergamon Press
NEW YORK • OXFORD • TORONTO • SYDNEY • FRANKFURT • PARIS

Pergamon Press Offices:

U.S.A.

Pergamon Press Inc., Maxwell House, Fairview Park, Elmsford, New York 10523, U.S.A.

U.K.

Pergamon Press Ltd., Headington Hill Hall, Oxford OX3 0BW, England

CANADA

Pergamon of Canada Ltd., 150 Consumers Road, Willowdale, Ontario M2J 1P9, Canada

AUSTRALIA

Pergamon Press (Aust) Pty. Ltd., P O Box 544, Potts Point, NSW 2011, Australia

FRANCE

Pergamon Press SARL, 24 rue des Ecoles, 75240 Paris, Cedex 05, France

FEDERAL REPUBLIC OF GERMANY

Pergamon Press GmbH, 6242 Kronberg/Taunus, Pferdstrasse 1, Federal Republic of Germany

Library of Congress Cataloging in Publication Data
Diwan, Romesh, 1933-
    Alternative development strategies and appropriate technology.

    (Pergamon policy studies)
    Bibliography: p.
    Includes index.
    1.   Underdeveloped areas.   2.   Economic development.
3.   Underdeveloped areas—Science.   4.   Underdeveloped
areas—Technology.   5.   International economic relations.
I.   Livingston, Dennis, joint author.   II.   Title.
HC59.7.D55   1979   301.24'3'091724   79-9839
ISBN 0-08-023891-2

Printed in the United States of America

To our parents:

Leela Devi and Fateh Chand Diwan
Ruth and Jerry Livingston

Contents

Preface

The idea of this book arose from a proposal titled "Development Strategies and Technological Choices in Developing Countries," funded by the National Science Foundation (from February to September 1978) to help the United States Department of State formulate policies toward the United Nations Conference on Science and Technology for Development (UNCSTED) scheduled for August 1979 in Vienna. This funding allowed us to organize a workshop in July 1978 at Rensselaer Polytechnic Institute, Troy, New York to specifically discuss our preliminary paper, and we submitted the final report on September 30, 1978. The major part of our report is incorporated into this book which is much larger and broader in scope.

Both of us have read and discussed the entire book and we agree on all the ideas and issues included here. For operational purposes, we divided our responsibilities. Romesh Diwan had primary responsibility for Chapters 2 through 6, 8 through 10, 12, and 15 through 17. Dennis Livingston wrote Chapters 1, 7, 11, 13, and 14.

The ideas in this book are based on the following assumptions:

1. People, even those who are poor, illiterate, and unemployed, are intelligent. They are capable of defining their own needs and given opportunities, they can and will solve their own problems.

2. Development means the development of each and every human being. This involves the satisfaction of basic needs, both material and nonmaterial.

3. Such development is both possible and feasible. Indeed, it is the imperative of the historical processes that started with decolonization.

4. Such development is relevant not only to developing countries but to developed countries.

Ours is a consensus model. We are impressed by the mutuality of interests between the developed and developing countries. Conflicts do exist, and bad and wrong decisions are made. They flow from the emphasis on short-term private interests and from paths that once seemed genuinely attractive. Conflicts develop from inequalities — within and among countries, from processes leading toward such inequalities, and from evolution and institutionalization of "privileges." However, we do not believe that human beings in general, and decision-makers in particular, are evil. Therefore our approach is integrative, of human beings and societies. It is integrative because it recognizes inequalities and concentrates on processes that reduce them. Our solutions are long-term, and will involve a number of generations. This does not mean that we do not provide short-term policies — they are implied in the long-term solutions.

The book is about the world and its development. Part I describes the international world order as it exists today, the international actors, international relations, and international resource constraints. It describes the existing typology of the world in terms of development.

The world is divided into four groups: developed countries, high-income developing countries, high technology developing countries, and other developing countries.

Part II looks at the development path that has been followed in both developed and developing countries. After independence and during the fifties and sixties, most of the third world countries opted for what may be called the Conventional Development Strategy (CDS). It has been based on two basic propositions, that the developed countries of the West provide an example of the "desired state of development," and that the strategy and policies based on the maximization of the rate of growth of GNP per capita lead to the "desired state of development." This strategy has also been followed in the developed countries and it defines a particular type of industrial science and technology which has evolved in the developed Western countries. Such science and technology are capable of mass-scale production and are capital, energy, resource, and skill intensive.

Under CDS, such science and technology has been transferred from the rich to the elite of poor countries. By now, CDS has been in operation in most developing countries for more than 30 years. There are a few success stories of countries that have achieved high GNP growth. Though CDS has created a general awareness and some pockets of development, on the whole, the "desired state" has been elusive. Quite a few problems, many unforeseen, have resulted. For example, the absolute number, as well as the proportion, of poor and unemployed has greatly increased; income inequalities have increased manifold; and in many countries dual societies of a rich elite and a poor majority have emerged. These problems need immediate attention. Furthermore, in terms of international relations, CDS has fostered, or at least encouraged, a dependence by the poor on the rich countries for basic, consumer, and capital goods. A rethinking about development strategies is needed, including within developed countries themselves.

Part III deals with Alternative Development Strategies (ADS).

Basic objectives of ADS are reductions in, and eventual elimination of, poverty, unemployment, underemployment, and income inequalities. The emphases are on the production of goods for basic needs of the poor, by the people themselves; and on local initiatives, resources, and self-reliance — not on GNP growth. Operationally, there is no one particular ADS. Instead, there are many variants of it. The International Labor Office variant emphasizesa "basic-needs strategy" for direct attack on the needs of the poor and unemployed. The Sixth Five-Year Plan of India talks of Gandhian development strategy and has defined a minimum needs program. The Arusha declaration of Tanzania outlines development through the "Ujamaa Village;" in Guinea-Bissau, the emphasis is on agricultural production and "political" literacy. In virtually all other countries the policies to achieve some of the objectives of ADS are being pursued.

While conventional, large-scale technology need not be automatically excluded by countries following ADS, the science and technology most closely associated with ADS are best described as appropriate technology (AT). AT is by the people themselves; location specific, holistic and future oriented. Based on the criterion and location of decision-making, various types of ATs are identified, e.g., family-employing and community level technology. Since these ATs match resources directly to objectives, they are generally efficient and cost least. Criteria for the choices of these ATs may be derived from the objectives of ADS and the characteristics of ATs.

ADS are of interest to developed as well as developing countries; they are relevant to the former as they encounter facets of overdevelopment that appear to be mitigating historical rates of growth. In their future development, developing countries will require ADS and concomitant technological processes heavily dependent on AT. Thus, rich and poor countries begin to share a common dialogue on alternative development paths and in this sense, ADS and AT encourage a relationship of genuine interdependence.

In Part IV policies consistent with ADS are derived. Some policies for developed countries and international agencies are outlined. The relevance of the New International Economic Order in the North-South dialogue is analyzed in terms of human needs. Areas for fruitful cooperation between developing countries, based on AT and ADS institutions, are suggested.

By dividing the developing countries among three groups, high income, high technology, and others; and by distinguishing within a country dual societies made up of a rich elite and poor majority, we provide sharper tools of analysis. The latter distinction is particularly important because treating a country as a unit of analysis suffers from the fallacy of "gross categories."

We are not unaware of the long struggle humanity has to go through to achieve an equitable world order, but we remain optimistic about the future of humankind. Hopefully this optimism is reflected in the book and provides hope and help in this long struggle.

Acknowledgments

In the process of completing this book we have talked to a large number of people. Some of them have taken a great interest and have participated in a workshop organized in July 1978 to discuss an earlier draft. Others have reviewed various parts and made many suggestions. However, we remain responsible for all the errors.

We would like to express our thanks to Chandra Agrawal (University of Michigan); Karim Ahmed (National Resources Defense Council); Chris Ahrens (Community Service Administration); Bill Anderson and Carolyn Rhodes (Congressman Long's office); Marianne Balchin; Harriet Barlow (Institute of Local Self-Reliance); Ann Becker and Jim Benson (Council on Economic Priorities); Ken Bossing (Citizen's Energy Project); David Chu and Ward Morehouse (Council on International and Public Affairs); Ken Dahlberg (Western Michigan University; Bill Eilers (A.I.D.); Bill Ellis (TRANET); David Felix (Washington University); Sushila Gidwani (Manhattan Colle); David Goldberg (Program Planning/Development Associates); Denis Goulet (Overseas Development Council); W.A. Gross (University of Mexico); Jim Gudaitis and Peter Henriot (Center for Concern); Jeff Liss (N.I.H.); Suren Navlakha (Institute of Economic Growth); Denis Hayes, Colin Norman, and Bruce Stokes (Worldwatch Institute): Gordon Hiebert, Aaron Segal, and Lynn Preston (National Science Foundation); S.R. Hiremath (India Development Service); Devinder Kumar (Wordha, India); Patricia Kutzner (World Hunger Education Service); Thomas Maloney (Southern Illinois University); Eleonara Masini (World Futures Studies Federation); Bagich Minhas and Chuck Weiss (World Bank); Madan Handa and Joya Sen (Ontario Institute for Studies in Education); David Pimentel (Cornell University); Jack Purdum (Butler University); Howard Osborne (USDA); Ted Owens (A.T.I.); McGregor Reid (Jet Propulsion Laboratories); Barbard Reno (C.U.N.A.); Lu Rudel (U.S. Department of State); Martha Steger (Georgia Tech), Heather Tischbein (National Rural Center), Michaele Walsh (Rockefeller Brothers); Christopher Wright (O.T.A.); Rudy Yaksick (Center for Community Economic Development); Joyce

Diwan, Ed Fox, Ed Holstein, Deborah Johnson, Robert Resnick, Sal Restivo, Rick Worthington, and Alex Wynnyczuk (R.P.I.).

We are grateful to Kamelia Alavi, Brian Denison, Colleen Sweinhart, and Edie Wiarda for help in various chores, and to Kathy Keenan and particularly Betty Jean Kaufmann for her patience and kindness in working with us and typing the various drafts for this book. Renu Bawa has kindly provided immense help in the compilation of the index.

Support for part of this work by National Science Foundation Grant INT 78-08330 is gratefully acknowledged.

I

Existing International Order

1 Existing International Order

THE TRADITIONAL INTERNATIONAL SYSTEM

The dynamics of the development process are intimately connected with the structure and operation of international relations. This connection takes various forms. While we will explore the details in succeeding chapters, several primary links may be noted.

It is within the international political system that the definition of "development" itself is formulated. Development may be perceived as predominately involving economic growth, as measured by increases in gross national product (GNP) and other quantitative factors, or as involving improvements in both material and nonmaterial standards of living. The latter could include such cultural and institutional bench marks as the capacity of an area to provide its own food and shelter or the strengthening of a country's ability to screen imports of technology. But whatever the range of connotations that might be attached to "development," the conceptualization most likely to be adopted is that which most closely fits the experiences and interests of the most powerful nations and other international actors.

"Development," then, is never a neutral term, however precise and scientific the measurements used to express the progress toward it may seem. Prevailing notions of development reflect the ability of those nations most able to enforce their particular definition, for example, by the provision, or withholding, of aid. It is also striking that the division of the world into "developed" and "developing" countries itself tends to express the world view characteristic of the affluent nations. These terms invariably carry economic connotations, although societies which are less well off economically are not necessarily "underdeveloped" in culture or other criteria of civilization.

It is also within the international system that modes of cooperation among nations for the facilitation of development are established. Political realities play an important role in shaping the form and functions of development programs. This is all the more likely since

3

there is no one obvious or objective pattern of international interaction which is guaranteed to promote development, however defined. Although we believe that there is a range of preferred development strategies, a broad area of discretion remains regarding the institutional means used to implement any strategy. Therefore, powerful states tend to orient development programs and organizations in which they are involved to match their own global roles. And thus many arguments have broken out in recent years over weighted voting for developed members of multinational lending agencies, tying of loans or grants from a country to purchases within that country, and donor country restrictions on aid to industrial sectors which might compete with their equivalent exports or domestic concerns. Such disputes do not revolve solely around the proper definition of development.

In searching for an appropriate development strategy, then, we must take into account those aspects of the present international system which appear most relevant in affecting the contours of development. This is particularly the case since we will make an argument for development options which do not match those heretofore favored by the dominant nations.

An understanding of the present system must, in turn, be predicated on an appreciation of the differences between international relations as traditionally conceived and practiced, and the ways this system has evolved in our time.(1)

The actors traditionally recognized as legitimate participants in international relations have been nation-states. Arising out of a period of prolonged warfare, the classical state system took shape in the early seventeenth century. Ever since, the state has served as the basic ordering unit for international relations. Each state is coequal with a geographical portion of the earth falling within its boundaries. Under the formal rules of the system, every state is entitled to exercise supremacy over its internal affairs and is correspondingly responsible for not interfering in the domestic affairs of other states. This system presumes that it is relatively easy to distinguish between those events which are wholly internal or domestic and those which are foreign to a state. It is also a prevailing rule that vis-a-vis each other, states are legally equal, a principle codified in the one-state, one-vote provision applied by the United Nations Charter to the UN General Assembly.

Thus the state has been traditionally conceived of as a discrete entity and the only actor legitimately entitled to take part in international relations. These formal notions of the sovereign state and its place in international life coincided reasonably with the conditions of state interaction at the time such doctrines emerged in the seventeenth and eighteenth centuries. The European countries that comprised the core of the system shared a homogeneous cultural background and history. Means of influencing state behavior were generally limited to direct contact among countries via official government-to-government diplomacy, trade, and war. Otherwise, most citizens had little concern for the affairs of nations other than their own. Finally, available technology posed no threat to the high degree of autonomy states could maintain over their own affairs. In sum, at one time states could fairly claim to be actually and nominally sovereign bodies, while governments could claim the sole right, in theory and practice, to take part in international rule-making and resource allocation activities.

All these factors have dramatically changed since the nineteenth century. The nature of these changes and their implications for development may now be briefly surveyed.

INTERDEPENDENCE

The formal doctrines of sovereign equality of states and the exclusive jurisdiction of a state over its own territory remain as guiding norms of the international system. But whatever truths they once reflected have been undercut by the emergence of a global polity. The imagery of a "web" of interdependence is often used to express the nature of current international reality. An alternative metaphor is that the world has been "stitched" together thanks to the pace of technological advance. The latter has given rise to overlapping networks that facilitate international exchanges in communications and transportation, trade and currencies, tourism, and science and technology.

However visualized, interdependence often carries favorable connotations in American political rhetoric. It implies neighborliness, cooperation, the global village – a world where national borders are historic relics of less enlightened times. It also partakes of the same sense of inevitability as technological progress itself. Interdependence is the wave of the future; "managing interdependence" will take care of any problems that crop up from this trend.

However the concept of interdependence is more complex than it it might seem at first, and it must be separated from automatic judgments of its worth. As we shall see, there are costs and benefits to interdependence which fall differentially on various states, and on social groups within states. Three patterns of interdependence may be distinguished within which development policies are formulated and implemented.(2)

First, interdependence implies mutual reliance. States must rely or depend on each other for the provision of wanted goods and services. These may be material, such as products exchanged in trade, or informational, such as ideological beliefs and scientific data that flow among nations. In this way, interdependence is actually a condition of shared dependencies. Thus developed countries depend on developing countries for a large proportion of their petroleum requirements and a variety of other resources, while developing countries are heavily dependent on affluent nations for many industrial products and technologies. The ideology of progress through technology-fueled growth has spread from Western countries throughout the world, while echoes of the strategies and slogans of Third World liberation movements have occured in the West.

Second, and following directly from the above, interdependence implies that issues of concern to states on the global political agenda are thoroughly intertwined. Political demands are interlinked, environmental reforms raise economic questions, resource shortages carry military implications, human rights violations call into question economic aid, implementation of energy conservation impacts balances of payment, and so forth. Development strategies are no exception. They also intersect with the range of issues just noted. Thus at the Law of the Sea Conference, an ongoing series of negotiations, developing countrys' concerns over distribution of mineral wealth garnered from the ocean floor has been caught up with a host of larger military, economic, and environmental problems, making progress on the seabed issue alone singularly difficult.

The result is a world of shifting alliances and bargaining patterns, depending on the issue and the stakes involved. Older alliances built around cold war politics have obviously not disappeared. Rather, issues of superpower dominance within the East-West rivalry have been supplemented by ad hoc coalitions which cut across the traditional concerns of the United States and Soviet blocs. The same may be said about the developed/ developing nations, or North-South, split. For some purposes, each grouping may be discussed as a unit, but in other respects (as the next chapter explains) the developing world has significant differences within itself.

Development questions can become hostage to the resolution or amelioration of related concerns. The place of development itself on the world agenda becomes part of the political bargaining process; whether or not developed countries give it the attention and resources that developing nations wish depends on the skill and leverage available to the latter. Development automatically rates first priority in international forums no more than any other issue. Moreover, different development strategies may win the support of different groups of states, located within both North and South. Defining a preferred development path is difficult enough, but this aspect of interdependence further embroils the task.

A third connotation of interdependence is the ability of states and other international actors to interpenetrate each other. Significant decisions or events originating wholly within national borders may have effects far from their point of origin and may, in turn, be influenced by equivalent occurrences not under a nation's immediate control. Thus national activities that purposefully or inadvertently alter the environment potentially impact the welfare of distant countries (for example, the flow of sulfur dioxide emitted from industrial processes across the borders of Western Europe and North America). Such transnational phenomena also occur in a wide variety of areas subject to national decision-making, including trade, immigration, inflation control, cultural exchange, and so forth. In view of the mutual vulnerability of states to each other's actions, any absolute distinctions between "domestic" and "international" affairs have become obsolete.

It is also important to note that governments are influenced by the efforts of subnational, as well as international, groups. Thus development policy is not the sole creation of government elites. In a developed country, concerned citizens groups may attempt to lobby officials regarding particular programs that affect their own economic well being (e.g., by encouraging foreign business competition) or sense of global justice. In a developing country, the government may be subject to pressure from international lending agencies to adopt more stringent monetary policies as a precondition of aid. Domestic groups may also work directly to alter the policies of international bodies or may enlist the support of sympathetic portions of their own government against other governmental branches. A nation's development policy, then, is vulnerable to a range of influences as it comes under the scrutiny of subnational political groups as well as other interested international parties.

INTERNATIONAL DISPARITIES

In the classical international system, the attributes of national power were usually defined by military criteria. In this sense, a rough parity existed among states, at least to the extent that no one of them had the ability to annihilate the rest. While war-making ability is still an important gauge of a nation's relative power status, the capacity of a state to influence the behavior of other states is also based on its skill in manipulating the facets of interdependence noted above. This could entail, for example, directing trade and aid flows to political clients, making appeals to foreign nationals who share ethnic bonds with the country's own, and lining up votes for favorable policy declarations in international assemblies. For these purposes, economic and techno-logical attributes are important; the following chapter uses several such indicators in categorizing levels of development.

By either military or nonmilitary criteria, in any case, enormous disparities clearly exist among the wide number of states admitted to membership in the present system. Thus 95 percent of the world's research and development takes place in the industrialized countries, while the bulk of the remainder is performed in developing nations by subsidiaries of corporations headquartered in the former. Some two-thirds of world trade and commodity exchange also takes place among the affluent countries, while developing nations contribute only 8 percent of the world's manufactured goods. Similarly, large proportions of world energy generation and use, newspaper circulation, automobiles per capita, and other totems of technological power are concentrated or originate in developed countries.(3)

The economic and social consequences of such disparities have been expressed in a familiar metaphor: "If the world were a global village of 100 people, one-third of them would be rich or of moderate income, two-thirds would be poor. Of the 100 residents, 47 would be unable to read, and only one would have a college education. About 35 would be suffering from hunger and malnutrition, at least half would be homeless or living in substandard housing."(4) It is these consequences, of course, which have provided a moral and political spark to the search for ways to narrow the material disparities between rich and poor countries.

The North-South gap is itself a reality of the political system which must influence the choice of both development strategies and means of implementing them. Most obviously, patterns of "mutual reliance" may mask conditions of asymmetrical relationships among groups of nations. Existing political, economic, and technological networks which interlace international affairs act to skew the range of development options open to developing countries toward those most favorable to the interests of the powerful. Moreover, countries which are already materially wealthy have the means to gain continuing and disproportionate access to the stock of world resources. These include not only minerals, food, and energy, but intellectual and managerial skills. Characteristically, the flow of such resources within the global community is determined more by purchasing ability and other attributes of power than by basic need. In addition, the guidelines by which resources are exploited, transferred, and traded are determined by international institutions whose decision-making processes are biased in favor of the developed countries' world view.

It has not escaped the notice of developing nations that states may be legally equal, but in reality some are more equal than others. Or to return to the earlier metaphor, if interdependence is a web, somewhere there is a spider. As a result of this perception, the fact of disparities has been pushed onto the world political agenda by developing countries in the form of an all-embracing program called the New International Economic Order (NIEO). Basically this concept stems from a moral imperative to alter those aspects of the international system which systematically work against the self-defined best interests of developing countries, thereby freezing them into an inequitable and intolerable status quo. Components of the NIEO include garnering a larger share of world manufacturing for developing countries, easing the access of such products to markets in developed countries, initiating a moratorium on developing country debts, providing for the automatic transfer of monetary resources from the rich to poor (such as through a tax on seabed minerals), increasing concessional assistance, easing patent restrictions on availability of technology, and stabilizing commodity prices. The presumption is that such reforms would enhance the economic well-being of developing countries and their participation in international decision-making forums.

The NIEO, then, has become an important focus for development strategies in developing countries. For their part, developed countries have been reluctant so far to make serious concessions in this direction. This is not surprising in view of the implied demands for a global shift in political and economic power that the NIEO represents. Developed countries have also raised the issue of precisely who would benefit within developing nations from such changes – governing elites or the poor majority? We will explore these themes in later chapter, in particular the relationship of the NIEO, a proposed solution to global disparities, to appropriate development paths.

INTERNATIONAL ACTORS

Traditionally, only states were entitled to take part in international processes of rule-making and other formal intergovernmental exchanges. Now, however, states are not the only actors in the international arena. At the global level, the state must compete and cooperate with other institutions for resources, prestige, and influence. This does not imply that the state is obsolete or unnecessary, but only that the range of entities with which it may potentially interact has expanded.

One set of actors comprises international organizations. Their numbers have greatly proliferated in the twentieth century in response to the need for institutionalizing and regulating the wide variety of international exchanges among states and their citizens. Intergovernmental organizations have formal status in the international system through their constitutive treaties which specify their legal rights and obligations. Such bodies provide valuable meeting grounds and conference facilities within which states can carry out, or not carry out, negotiations on which issues should receive priority on the global agenda and how to implement them. Indeed, international organs based on the one state/one vote principle have given a voice to the demands of developing countries that would be otherwise unavailable. The resolutions emerging from these groups are not of a binding character. If backed by sufficiently large voting majorities, they may portend the emergence of new rules and value shifts in the international system. Moreover, some organizations have become important agencies in their own right because of the mediation services and financial aid they can provide members.

The NIEO and development options are among the issues that have been under intense debate within a number of intergovernmental organizations. The more significant of these are located within the United Nations family of specialized agencies and subsidiary organs, including the International Bank for Reconstruction and Development (World Bank), UN Development Program (UNDP), UN Environment Program (UNEP), UN Industrial Development Organization (UNIDO), International Labor Organization (ILO), and UN Conference on Trade and Development (UNCTAD). Periodic ad hoc conferences, such as the UN Conference on Science and Technology for Development (UNCSTD), are important in this respect, as are such regional groups as the European Economic Commission, the Organization of Petroleum Exporting Countries (OPEC), and the Organization for Economic Cooperation and Development (OECD).

There also exist an even larger number of nongovernmental, or private, international organizations. Those active in development include the Society for International Development, World Council of Churches, and International Coalition for Development Alternatives. Such groups can take on lobbying functions vis-a-vis intergovernmental organizations equivalent to the role of public interest groups within states and provide expertise, ideas, and information on relevant activities that members can use in attempts to influence development programs within their own countries.

Besides international organizations, a second set of actors that has risen to prominence is the multinational (or, in UN parlance, transnational) corporation. A large literature exists comparing the power and impact of the multinationals, favorably and otherwise, to states themselves.(5) They are, in a sense, miniature world governments, recruiting personnel, exploiting resources, and selling products on a global basis that transcends the more limited capabilities of most countries. In the development process their role is crucial and controversial. Viewed positively they act as agents for the transfer of advanced technology and associated skills from developed to developing countries, enabling the latter to enjoy the fruits of technological progress and to enter world trade markets at a faster pace than would otherwise be possible. Their opponents see them as oriented, by definition, to the perpetuation of their own interests, independent of the needs of developing or developed countries and not really subject to the latter's control. The location of the home base of most multinationals in industrialized nations also feeds suspicion that their presumptively global outlook actually meshes with and supports the foreign policies of Western countries. Thus we will be concerned with the role of multinations in facilitating or hindering the implementation of preferred development strategies.

VIOLENCE AND REPRESSION

As noted above, the role of force in the modern world has taken on a more complex role than in the classical system. In the latter, war was accepted as a rational, legitimate last resort (and often a less-than-last resort) for resolving disputes. The strength of a state was closely identified with its military capabilities. Now the dangers of escalation have made traditional forms of warfare among states possessing nuclear weapons if not unthinkable, at least inconceivable as a rational strategy for any goals short of retaliation for prior atomic attack. War is also too clumsy or inappropriate a tool for the range of economic and technological objectives states seek in an interdependent world. For the same reason, many methods short of force are available to bring pressure on a state's bargaining partners, including the promise of aid or threat to withdraw it, suspension of cultural exchanges, and the like. In legal terms, the UN Charter obligates its members, in any case, to refrain from the use or threat of force against each other.

However, none of the above has prevented states from engaging in a never-ending and debilitating global arms race. The super powers continue to stockpile expensive weapons systems in the name of mutual deterrence, where the possibility of parity is never able to overcome the desire for superiority. Developing countries import large quantities of conventional weapons from armaments manufacturers in the developed world for use against each other. The only beneficiaries are the weapons industries, whose global output is approaching the astonishing sum of $400 billion per year.(6)

One obvious implication of such spending for development is the burden it puts on the already limited budgets of developed, and even developing, countries. All states must make trade-offs between available resources and burgeoning demands for various national programs, but governments of most developing countries face this problem in acute form. There is a cruel dilemma here. To some extent the tensions among poor nations that stimulate military spending are themselves rooted in the corollaries of underdevelopment (low per capita income, energy use, etc.), which in turn become more difficult to deal with when resources are diverted to military needs. The same connection holds true, perhaps to a greater degree, under the presumption that a major use of military forces for developing country governments is to suppress internal dissension arising from rich-poor disparities within these nations, in turn made worse by the creation of large standing armies. We are not arguing that simply increasing the sums allocated to development projects is the best path to take. Much of this book is devoted to the need for conceptual, political, and institutional changes in traditional development plans, independent of funding. But monetary resources are clearly needed somewhere along the line; local and regional arms races provide a social black hole for their use.

A more subtle effect of preoccupation with military affairs is the diversion of intellectual resources to such maladaptive purposes. Scientific skills, engineering know how, managerial abilities, and the like are all scarce resources as much as money. It is impossible to judge what contributions might be made to development if the human talent bound up with military pursuits could be converted to civilian enterprises. The suspicion lingers that it would be great, especially if harnessed to the strategies we suggest.

Apart from the effects of military spending per se, the armed forces available to the developed countries always carry the potential for interventionary use in developing nations. This is particularly the case for countries that implement strategies for economic growth perceived as threatening to the national security of industrialized states. Speculation was rife about the United States sending marines to take over Middle Eastern oil fields during the 1973 oil embargo. Short of military force, developed countries in the East and West possess extensive capacity for more covert manipulation of government policies, or changes in governments themselves, in developing countries. Depending on the nature of the situation, this factor may impose a serious, though not insuperable, restraint on the elaboration of alternative development strategies by developing country officials.

In sum, as one observer has put it, "development is peace – peace is another word for development."(7) This accurate perception requires that we pay heed to the relationships between development paths and national capacities and predilections for internal and international violence.

ECOLOGICAL CONSTRAINTS

Questions of resource availability, access, and use bear directly on prospects for development in our time. Under the classical international system, ecological issues in the broad sense could be ignored. Well into the beginnings of the industrial age, resources were relatively easy to obtain and the ability of the environment to absorb the burden of pollutants was not yet, in most areas, overstrained. This is no longer the case. Environmental resource concerns have risen high on the international agenda.

We see no need to rehearse in detail yet another account of the limits to growth debate. It is sufficient to say that we believe legitimate questions of global interest have been posed around the "world problematique." This covers such issues as pervasive, cumulative, and long-lasting environmental pollution; a tumbling of the international monetary system; large price increases in the cost of gasoline and other sources of energy; dwindling food reserves; persistent, simultaneous inflation and unemployment in industrialized countries; and the difficulties of large bureaucratic institutions in managing the complexities of modern life.(8)

The major point here is not that there may or may not be absolute physical limits to resource availability, but that there are restraints on the ability of nations to continue unlimited growth patterns. These restraints are not only environmental, but social, political, and psychological. There are many technological "solutions" to breaking through energy and other scarcity barriers, but they involve costs which are increasingly unaffordable by even the richest countries, in terms of availability of equity capital, difficulty of obtaining remaining resources, and widespread disaffection from the material benefits of growth. In short, the opportunity costs of continuing accelerating growth may have exceeded the productivity gains resulting from such growth.(9)

These considerations must influence the search for development options which will not lead to dead ends or increased social strife. At minimun, they imply that the implicit goal of many developing countries to achieve the current standard of living and industrial infrastructure of the developed nations may be a dream whose time has passed – for developed, as well as developing, countries.

2 Development: A World Typology

INTRODUCTION

Looked at from the moon, the earth looks like one compact ball. For purposes of analysis, however, the world is too large. It needs to be divided and subdivided into more homogeneous units. Economists have divided the world into two units: the developed or the rich countries and the underdeveloped or the poor countries. The United Nations, on the other hand, has made another category – socialist economies. The world is thus divided into developed market economies, socialist economies, and developing economies. The socialist economies contain both developed and developing countries. The reasons for making a separate category of socialist economies is that these countries are not easily comparable with countries based on market economies. Their statistics are different; they talk of material balance instead of GNP. From the science and technology point of view, on the other hand, one finds quite a few similarities between developed market economies and developed socialist economies.

DEVELOPED VERSUS DEVELOPING COUNTRIES: GENERAL COMMENTS

The developed market economies form a very small part of the world. These are approximately 20 to 25 countries represented in O.E.C.D. These have some common characteristics. A majority have per capita income levels of U.S. $2,000 or above. They have well-developed markets. A large section of the economy is employed in industrial production. Labor productivity is high, and in the last 15 years (1960 to 1975) it increased by 70 percent, while industrial production doubled.(1) They produce their own technology. In view of similarities, technology invented and defined in any of these countries finds an immediate use and relevance in other countries. The trade among these countries is large; the exports of developed countries to each other form 74 percent of their total exports.(2) Even in food and agricultural production these countries are well off.(3) They are committed to the goal of full employment. In recent years, a majority of them have suffered from the problems of stagflation – a simultaneous existence of stagnation, implying unemployment, and inflation. Stagflation is apparently becoming persistent. It poses a serious threat to these countries and to their political stability. In addition, there is now a keen competition among some of these countries, thereby straining relationships between them (for example, Japan and the United States.) They are facing some serious problems arising from the increase in the price of oil by O.P.E.C. This has already affected arrangements in the International Monetary Fund. The status of the U.S. dollar as a medium of international exchange is under threat. In addition, some of the externalities of technological progress, such as pollution, environmental degradation, and health risks, have become persistent. Some of these problems reveal facets of overdevelopment which are generating questions about virtually all aspects of the economy, particularly the costs and benefits of science and technology, hard versus soft paths, quality of life, and the like.

If developed market economies define a small and relatively homogeneous part of our world, developing countries(4) describe a large and heterogeneous part. For any useful purpose, it is necessary to divide the world of developing countries still further. At the end of the Second World War, a number of these countries were colonies of some of the countries who now constitute the developed part of the world. However, as these countries have gained independence, the relations between the colonial country and the erswhile colonies have weakened. Commonwealth countries associated with the British empire, the union of French-speaking African countries formerly ruled by France, and the Organization of American States still reflect some of these relations. For analytical purposes, these developing countries are now divided by the level of development.

Generally, developing countries are described as developing in relation to other countries which are considered developed. Developing countries are considered to be in the process of development. Development is variously described. Developing countries are those who do not have a very high per capita income, where the infrastructure in terms of communications and service is still being developed, and where a large part of the population and the economy is dependent upon agriculture, a particular crop or group of crops, or a few raw materials. In spite of this, the developing countries are hardly self-sufficient in food production.(5) The labor productivity, particularly in manufacturing, is low.(6) There are few opportunities for children to go to school, few colleges and universities, few technologically trained people. These countries import technology and have to depend upon the rich (advanced) countries for technological innovations, and for other goods. Their major trade is with the developed countries. In spite of so many developing countries, trade with each other is limited; it is less than one quarter of total world trade.(7)

TYPOLOGY OF DEVELOPMENT

From the point of view of our interest in science and technology (S&T henceforth), all countries of the world may be divided on the basis of per capita incomes, which is considered one if not the only indicator of development; and the level of science and technology development. In terms of these criteria, there are large variations within the developing countries. Some of the developing countries have low per capita incomes, others have relatively high incomes. Some developing countries have certain well-established sectors while others do not.

If we look at the economists' criteria of development in terms of per capita incomes, we can divide the world into four categories. Table 2.1 below provides information about which countries belong where. The table has been derived from the Overseas Development Council's data on developing and developed countries. There is no well-defined per capita income that will sort the rich from the poor; development, after all, is not only a matter of per capita income. The classification, then, remains arbitrary.(8) Our purpose here is to develop a stylized typology of world development which is not affected by a particular country being in one or the other class. Even though our classification criteria are somewhat arbitrary, they do bring out the pattern generally associated with developed and developing countries.

Looking in the "high" class we find two groups of countries, those that are generally associated with developed countries, and the oil-exporting countries of the Middle East. Generally speaking, these countries are not considered developed even though in terms of per capita income they are rich countries; as rich as the rich developed countries.

There is no way of defining the level of science and technology in a country. It depends upon a number of factors, such as stocks of scientists, engineers, and other skilled personnel, R&D expenditure, the nature of industrial activity, and so forth. Not only is the definition difficult, but so is the data. In many countries such data do not exist. In view of these difficulties, we suggest the following three indicators: the number of scientists and engineers in a country; the number of students enrolled in colleges and universities in a country; and the number of industrial units in production. Obviously these are rough indicators, but they do, hopefully, catch the pattern we are after. The number of scientists and engineers reflect that a country has a stock of technical expertise. It is possible that some of these scientists and engineers may not be exactly employed in science and technology; some of them may even be unemployed. Similarly, the number of students in colleges and universities reflect the potential of skill formation, but it is again possible that many may be students in humanities and arts. Similarly, the number of industrial units reflect the level of technological tasks, yet they may be obtained on the basis of different definitions. In other words, we recognize the weakness of the data. But they are, perhaps, the only data available. They are certainly helpful in defining a stylized profile of development.(9)

Table 2.2 below groups the various countries in four classes on the

Table 2.1. Grouping of Countries by Per Capita Income
(in U.S. dollars)

Group (and Criteria)	Number of Countries	Name of Countries
Low ($300 and less)	48	Afghanistan, Bangladesh, Benin (Dahomey), Bhutan, Bolivia, Botswana, Burma, Burundi, Central African Rep., Chad, Comoro Is., Egypt, Equatorial Guinea, Ethiopia, Gambia, Guinea, Haiti, India, Indonesia, Kenya, Khmer Rep., Laos, Lesotho, Macao, Malagasy Rep., Malawi, Maldive Is., Mali, Mauritania, Nepal, Niger, Nigeria, Pakistan, Rwanda, Sierra Leone, Sikkim, Somalia, Sri Lanka, Sudan, Tanzania, Togo, Tonga, Uganda, Upper Volta, Vietnam, Socialist Rep. of, Yemen Arab Rep., Yemen People's Rep., Zaire.
Lower Middle ($300 to $900)	39	Albania, Cameroon, Cape Verde Is., China People's Rep., Columbia, Congo People's Rep., Cuba, Dominican Rep., Ecuador, El Salvador, Ghana, Grenada, Guatemala, Guinea-Bissau, Guyana, Honduras, Ivory Coast, Jordan, Korea, Dem. People's Rep., Korea Rep. of, Liberia, Malaysia, Mauritius, Mongolia, Morocco, Mozambique, Nicaragua, Papua New Guinea, Philippines, Rhodesia, Sao Tome & Principe, Senegal, Swaziland, Syria, Thailand, Tunisia, Western Samoa, Zambia.
Upper Middle ($700 to $2,000)	35	Algeria, Angola, Argentina, Barbados, Brazil, Bulgaria, Chile, Costa Rica, Cyprus, Fiji, Gabon, Guadeloupe, Hong Kong, Iran, Iraq, Jamaica, Lebanon, Malta, Martinique, Mexico, Netherland Antilles, Oman, Panama, Peru, Portugal, Reunion, Romania, South Africa, Surinam, Taiwan, Trinidad and Tobago, Turkey, Uruguay, Venezuela, Yugoslavia.

| High ($2,000 and up) | 37 | Australia, Austria, Bahamas, Bahrain, Belgium, Canada, Czechoslovakia, Denmark, Finland, France, German Dem. Rep., German Fed. Rep., Greece, Hungary, Iceland, Ireland, Israel, Italy, Japan, Kuwait, Libya, Luxembourg, Netherlands, New Zealand, Norway, Poland, Puerto Rico, Qatar, Saudi Arabia, Singapore, Spain, Sweden, Switzerland, USSR, United Arab Emirates, United Kingdom, United States. |

Note: This table has been derived from the larger table in the Appendix.

basis of their stock of scientists and engineers. It will be noticed that data are available for only 67 countries. (It is very difficult to obtain.) Our classification is again arbitrary, but it brings out the level of S&T in the country. Once again, looking at the "high" group, we notice that there are two types of countries: the developed countries and some large developing countries such as India, Argentina, Brazil.

Table 2.3 below provides a similar classification on the basis of enrollment in colleges and universities. In this case we have data for 13 countries. The classification is arbitrary, and the "high" class contains both developed economies and a group of large developing countries.

In Table 2.4 below the countries are classified by the number of industrial units reporting. Here the data are for 102 countries, but the classification is also arbitrary. The "high" class contains developed economies and some large developing economies.

Since the countries in the developing world are classified in one or the other "high" class, we can sort out those developing countries that belong to a "high" class by at least two criteria.

WORLD DEVELOPMENT: A STYLIZED TYPOLOGY

On the basis of the tables and for purposes of our analysis, the world may be divided into the following four categories:

1) Developed Countries: These are the countries with high levels of incomes and S&T. We have already commented about them.

2) High Income Developing Countries: In our classification, these are the countries with per capita income of more than

$2000 in 1976. The number of countries in this group is a small one. For all intents and purposes, these are oil-exporting Middle Eastern countries, namely Bahrain, Kuwait, Libya, Qatar, Saudi Arabia, and United Arab Emirates. Among these, Kuwait, Libya, and Saudi Arabia are the most important. For all analytical and policy purposes, it is these three countries that need to be considered.

3) High Technology Developing Countries: These are the large countries that have developed a certain level of S&T infrastructure (8 to 15 countries depending upon particular classificatory criterion). The most important 8 are: Argentina, Brazil, and Mexico; India, Indonesia, Iran, and Korea in Asia; and Egypt in Africa. The extended list will include Chile and Peru in South America; Pakistan, Philippines, Taiwan, and Syria in Asia; and Nigeria and Kenya in Africa. By and large, these countries have well-developed industrial sectors. They compete with the developed countries in the developing and developed markets for consumer goods and they have large populations (a large part ₒf which is rather poor.) The per capita incomes in India, Indonesia, and Egypt are low. In other countries, however, per capita incomes are in the middle level. The People's Republic of China has not been included in these classes because no data is available. If data were availabe, our guess is that China would fall in this category. However, the conditions in China are far different from the conditions in the other 8 countries.

4) Other Developing Countries: The other 110 to 120 countries belong to this class, which has large variations. A detailed analysis will make finer distinctions. However, for our purposes, all of these countries with varying levels of per capita income, levels of S&T, size, and populations constitute one category. These are mostly small countries who are on the receiving end both in terms of S&T and economic aid.

Obviously, the relations among these four groups will be very different. There are at least six relations based on one group dealing with the other. The relations among developed countries and high income developing countries and high technology developing countries are different. Similarly, high technology developing countries treat high income developing countries differently from other developing countries. These groups of countries have different interests and these interests become clear, or confused, in international conferences and negotiations. Different policies and institutions affect these groups differently.

Relations between the developed countries, on the one hand, and different categories of developing countries, on the other, will be rather different. We pursue this issue in Part IV. But first we will discuss how the developing countries have been attempting to develop themselves. In terms of development strategies the distinctions outlined above become more or less irrelevant. In this context, all of the developing countries can be treated as one group.

Table 2.2. Grouping of Countries by Number of Scientists and Engineers

Group (and Criteria)	Number of Countries	Name of Countries
High (41,000 and up)	30	Argentina, Australia, Austria, Belgium, Brazil, Bulgaria, Canada, Czechoslovakia, Egypt, Finland, France, German Fed. Rep., Greece, Hungary, India, Iran, Korea Rep. of, Mexico, Netherlands, Norway, Pakistan, Peru, Philippines, Poland, Romania, Spain, United Kingdom, USSR, United States, Yugoslavia.
Upper Middle (41,000 to 20,000)	10	Bangladesh, Denmark, Hong Kong, Ireland, Israel, Malaysia, Nigeria, Saudi Arabia, Singapore, Uruguay.
Lower Middle (20,000 to 4,000)	15	Afghanistan, Bolivia, Burma, Ghana Guatemala, Iraq, Jordan, Lebanon, Sri Lanka, Sudan, Tanzania, Thailand, Zaire, Zambia.
Low (4,000 and less)	12	Bahrain, Botswana, Cameroon, Fiji, Iceland, Kenya, Kuwait, Malta, Mauritius, Mongolia, Qatar, Togo.

Note: This table has been derived from the larger table in the Appendix.

Table 2.3. Grouping of Countries by Colleges and University Enrollment

Group (and Criteria)	Number of Countries	Name of Countries
High (100,000 and up)	39	Argentina, Australia, Bangladesh, Belgium, Brazil, Bulgaria, Canada, Chile, Czechoslovakia, Denmark, Egypt, Germany, German Fed. Rep., Hungary, India, Indonesia, Iran, Israel, Italy, Japan, Korea, Demo. People's Rep., Korea Rep. of, Mexico, Netherlands, New Zealand, Norway, Peru, Philippines, Poland, Puerto Rico, Romania, Spain, Sweden, Turkey, United Kingdom, USSR, United States, Venezuela, Yugoslavia.
Upper Middle (10,000-100,000)	39	Afghanistan, Albania, Algeria, Austria, Bolivia, Burma, Costa Rica, Cuba, Dominican Rep. Ecuador, El Salvador, Ethiopia, Finland, Greece, Guatemala, Guinea, Hong Kong, Iraq, Ireland, Kenya, Lebanon, Libya, Morocco, Nepal, Nicaragua, Nigeria, Portugal, Romania, Saudi Arabia, South Africa, Sri Lanka, Sudan, Switzerland, Syria, Thailand, Tunisia, Uruguay, Vietnam, Socialist Rep. of, Zaire.
Lower Middle (1,000-10,000)	30	Angola, Benin (Dahomey), Cameroon, Congo, Fiji, Ghana, Guinea, Guyana, Haiti, Iceland, Jamaica, Jordan, Khmer Rep., Kuwait, Malawi, Malaysia, Mali, Malta, Martinique, Papua New Guinea, Rwanda, Senegal, Tanzania, Togo, Trinidad and Tobago, Uganda, Zambia.

| Low
(1,000 and less) | 22 | Bahrain, Botswana, Burundi, Central African Rep., Chad, Cyprus Equatorial Guinea, Finland, Gabon, Gambia, Guadeloupe, Laos, Lesotha, Luxembourg, Niger, Reunion, Surinam, Upper Volta, Western Samoa, Yemen Arab Rep., Yemen People's Rep., United Arab Rep. |

Note: This table has been derived from the larger table in the Appendix.

Table 2.4. Grouping of Countries by Number of Industrial Units

Group (and Criteria)   Number of Countries

High (10,000 and up)	21	Australia, Belgium, Brazil, Canada, German Fed. Rep., Hong Kong, India, Indonesia, Israel, Japan, Korea Rep. of, Phillipines, Poland, Portugal, South Africa, Spain, Sweden, Syria, United Kingdom, USSR, United States.
Upper Middle (5,000-10,000)	15	Austria, Denmark, Finland, German Dem. Rep., Greece, Iran, Jordan, New Zealand, Norway, Peru, Switzerland, Thailand, Turkey, Uruguay, Venezuela.
Lower Middle (1,000-5,000)	28	Algeria, Angola, Bangaldesh, Bulgaria, Chile, Costa Rica, Cyprus, Dominican Rep., Ecuador, Egypt, Guatemala, Hungary, Iraq, Ireland, Jamaica, Kuwait, Malta, Mauritius, Morocco, Nigeria, Pakistan, Rodesia, Romania, Singapore, Sri Lanka, Tunesia, Yugoslavia.

| Low
(1,000 and less) | 38 | Afghanistan, Barbados, Bolivia, Botswana, Burundi, Cameroon, Central Africa Rep., Congo, Costa Rica, Czechoslovakia, El Salvador, Fiji, Gambia, Ghana, Guinea, Guyana, Haiti, Honduras, Kenya, Libya, Luxembourg, Malawi, Malaysia, Mali, Mongolia, Niger, Panama, Rwanda, Somalia, Swaziland, Tanzania, Togo, Trinidad and Tobago, Uganda, Upper Volta, Vietnam Socialist Rep. of, Yemen, Zambia. |

Note: This table has been derived from the larger table in the Appendix.

II
Conventional Development Strategy

3 A Historical Note

INTRODUCTION

The purpose of this chapter is to place the ideas about development in historical perspective. Historical processes are complex; much depends upon where and when the anaysis begins, what factors and events are highlighted, and which relationships are developed. There are, thus, a number of possible interpretations and explanations for virtually every phenomenon, idea, event. What is presented below is one of the explanations that seems to us to be the most plausible.

Prior to the Second World War, a large number of countries in the developing world were colonies of colonial countries in Western Europe. India, Pakistan, Bangladesh, Sri Lanka, Malaysia, Nigeria, Tanzania, Zambia, Kenya, Uganda, Sierra Leone, and Jamaica, to name only the larger ones, were the colonies of Britain. France colonized Algeria, Tunisia, Central African Republic, Chad, Mauritania, Niger, Haiti, Senegal, and many smaller countries like Martinique. Zaire was the Belgian colony of Congo, and Indonesia was a colony of Holland. Portuguese troops left Angola only a few years ago. In addition to outright colonies, the colonial powers exercised major control in many other countries such as Egypt, other Middle Eastern Arab countries, Ethiopia, and a large number of South American countries. Even though these were not exactly considered colonies, the influences had elements of a colonial relationship. The colonial relationship has existed for a few centuries, and it has left its impact. (Some argue that this influence continues even today.) To appreciate the current situation in the developing countries, it is helpful and even necessary to understand the colonial situation.

THE COLONIAL SITUATION

The colonial situation arises from brutal exploitation by the colonizer of the colonized. This exploitation extends to every aspect of the colonized person: body, mind, output, land, resources, home, family, language, history, tradition, and so forth. The colonizer obtains all privileges and the colonized is denied every right. This inequity is maintained by the force of the colonizer who possesses all power. Thus the colonizer's police and army is heavily armed, and means of force is denied to the colonized. The possession of a pocketknife can be a serious offense, sometimes punishable by death. The laws are designed to favor the colonizer; the colonized is always suspect, and can be arrested, maimed, and even killed on the flimsiest pretext. The colonial situation breeds, and is enforced by, violence. As Memmi and Fanon have persuasively argued, it affects both the colonizer and the colonized.(1)

The colonial situation gets rooted when the colonizer is able to force, by violence, and by acceptance of the colonized of what Memmi calls the "mythical portrait" of the colonized. The colonizer develops an image of the colonized, and in this image the differences between the colonizer and colonized are exaggerated.(2) The colonized is depicted as backward, coward, evil, lazy, sentimental, superstitious, thievish, unintelligent, virtually stupid, unscientific, weakling, wicked, and so forth. This image is not based on any existing reality, nor is it consistent. However, the image provides a justification for all the power and privileges that the colonizer obtains, and the excessive force that he uses. Obviously, a person with such traits needs to be beaten, protected, ruled, and saved. Such a person cannot live except in poverty. The colonizer thus does a favor by treating the colonized brutally. In the heydays of colonialism, such ideas were congealed into what was then called the "white man's burden."

Since such an image does not conform to reality, the colonizer sets up the task of creating a reality that fits this image. The colonial institutions, then are developed to create such colonized persons, i.e., to make the colonized behave in a manner that can be interpreted in the context of such an image. By continuous aggression and force the colonized is diverted, driven, and poked toward such behavior. When some colonized persons behave in this manner, the colonizer has created a semblance of reality reflecting this imaginary image of the colonized. The histories of the colonial era and colonized people are full of examples, experiences, and literature that provide ample evidence of this propostion.

RESPONSE OF COLONIZED: ASSIMILATION

The colonial situation impoverished the colonized; both the colonized person and the colonized as a people. In Memmi's words, "Everything in the colonized is deficient. Everything contributes to this deficiency."(3) The colonized is forced to adopt a behavior which at best is deforming; he can either reject or accept the ideology of the colonizer. Rejection of the ideology, however, is an act of revolution. Rebels are punished, heavily punished by death. Acceptance, on the other hand, can save one from outward violence. Actual behavior based on this acceptance may even bring small rewards.

The colonial ideology is based on two propositions. In its somewhat general form it may be expressed as follows. Everything associated with the colonizer is "good," and everything associated with the colonized is "bad."(4) Accepting this ideology, then, involves appreciating everything associated with the colonizer and rejecting everything associated with the colonized. The behavior based on this "acceptance" requires condemnation of everything the colonized is born into and acquisition of everything that seems to be associated with the colonizer. Thus, the colonized discards the traditional dress and adopts the dress of the colonizer. This is the simplest act. Similarly, the colonized adopts the food habits of the colonizer and rejects the food habits of the colonized society. As the level of acceptance increases, the colonized even speaks the language of the colonizer at home. Such activities, based on the acceptance of the colonial ideology, may be called the assimilative effect of colonialism.(5)

Under assimilation the colonized desired and acted to become "like" the colonizer. It took the form of imitation of the patterns of living and behavior of the colonizer. To do so effectively, the colonized had to learn the manners, language, and activities of the colonizer.   This implied that the children be sent to schools where the language of instruction was of the colonizer. The ideal in this respect was, of course, to go to the schools and colleges where the colonizers themselves obtained their education and training.(6) This meant going to study in the home country of the colonizers. But these activities also had a second dimension, a wholesale rejection and condemnation of the history, practices, traditions, and wisdom of the home society.(7) Such assimilative activities had their rewards.(8) However, the assimilators were not assimulated; they were not even accepted into the society of the colonizers. Race, skin, color, and accent, gave them away and kept them apart. And these native imitators of the colonizers were given junior, unpleasant roles in the administration of colonialism.

Assimilation is a dynamic process. We have derived the concept of assimilation from the colonial situation. It is one of the two choices open to the colonized. There are large variations within such assimilative behavior. However, this concept of assimilation is rather stringent and it is dependent on the colonial situation. It is possible to derive a broader concept of assimilation as well which follows from the existence of vast and "effective" inequities.(9) By effective we mean that the richer and powerful can heavily influence the poor and the weak.(10) In the broader sense, assimilative behavior takes place in situations which are not strictly colonial. Many authors contend that present relations between the United States and South American countries exhibit such "assimilative" influences. This is also true in the case of relations between such countries as Iran, Egypt, other Middle Eastern Arab countries, on the one hand, and France, England and the United States on the other.

NATIONALISM

The Second World War weakened the colonial countries in Western Europe, and accelerated the struggle by the colonized countries for freedom. The freedom struggles were intensely national and the parties fighting for freedom had broad based support from all groups. Also, after the Second World War, the United States emerged as one of the most powerful nations, and its policies were noncolonial, even anticolonial. The United States, thus, supported the national struggles for political freedom, thereby strengthening the hands of nonrevolutionary leaders. Within a decade after the Second World War, a number of countries gained independence; and the process accelerated so that by the mid-1960s virtually all of the colonized countries became nationalist independent states.(11) It is these countries that constitute a major part of the Third and Fourth World of today.

It is interesting, even important, to note that in gaining their independence, a majority of the colonized countries did not have to wage a long, protracted, violent struggle against the colonial government.(12) The Mau Mau in Kenya may be considered as a possible violent struggle, but this was quite short in time and more an aberration than a well-defined, long struggle. Algeria, Angola, Mozambique, and Vietnam are perhaps the only exceptions. The reason for the lack of violent struggles, in spite of the fact that a colonial situation is inherently violent, is obvious. No colonial system can fight and maintain its supremacy against the will of the people and their national struggles. France learned this lesson in Algeria at a very heavy cost. For the continuance of colonial rule and privilege, the success and prevalence of the "assimilation" principle is important. National, particularly violent, struggle weakens the "assimilation" process; the longer the struggle, the greater the elimination of assimilation. It takes time because assimilation is fundamentally a dynamic process; the longer the struggle, the greater the elimination of assimilation. It takes time because assimilation is fundamentally a dynamic process with a long gestation period and a long life of its own. On the other hand, if the national struggle is less violent and short, its impact on "assimilation" is very limited. Accordingly, after national independence in these countries the "assimilation" process continued unabated, and took different forms.

When these erstwhile colonized countries obtained independence, the governments were formed by nationalist parties. The leaders of these parties had accepted at least large parts of the colonial ideology. A number of these influential leaders had their education in the colonizer's home countries. With the government they took over all the colonial governing apparatus, with its associated mechanisms of repression against the opposition on the one hand, and privileges for those in control, on the other (for example, luxuriously furnished houses with servants, large posh offices with a hierarchy of bureaucrats and minions, and so forth.) To run this apparatus, they sought the help of native imitators of the colonizers. By and large, this administrative and bureaucratic structure is still intact in virtually every country. At best, minor changes have been made. The principle (the hierarchy, its insensitivity to suffering, and its irrelevance to the needs of the people) continues.

Part of the colonial ideology has implied that the colonized, hence the national leaders, are not capable of governing. The national leaders, accordingly, have been obsessed not only to govern but also to show internationally that they can govern. Maintenance of the national integrity of the State, therefore, became one of the two overriding issues. For example, the Gandhian idea of decentralization was flatly rejected.

The concern has been translated, year after year, into larger police and military budgets. It is no accident that the rate of growth of expenditures on police and military in these countries has been the highest. The developed economies have found a large and profitable market in military sales. It is no wonder that armaments have been the fastest growing industry in the world. The second effect is that these concerns and military expenditures have accentuated already existing suspicions among neighboring governments. The relations among newly independent countries have been less than cordial, and these have militated against cooperative efforts. (13)

The second overriding concern of these leaders was to bring these countries out of the mire of the past and into the modern world. In their view, the desired state and society was that of the colonial countries, with established and flourishing industries, urban centers full of cultural activities, universities and traditions of science and technology, "fast" life and new goods, and so forth. Their own countries, by comparison, were full of poor people living in villages, following old traditions and unexciting agricultural activities. These leaders, thus, defined the objective function of the government and the society to be modernization of itself in the image of the developed, and erstwhile colonial, countries.   Some leaders, such as Mahatma Gandhi, suggested other paths and objective functions involving improvements of the villages and the uplift of the poor and weak. This did not agree with the majority of the leaders in the government who had accepted parts of the colonial ideology. Nehru, prime minister of India from its independence to his death, reflected these attitudes when he said that large projects — such as steel plants, dams — were the temples of modern India.

IDEAS ON THE ROLE OF GOVERNMENT

In 1917 a revolutionary government took control of Soviet Russia. By the Second World War's end, it emerged as one of the important military powers. In the 1950s with Sputnik and the atom bomb, it took another leap and established itself as one of the super political powers. These seem to be major achievements for a period of 40 years or so. A number of leaders in developing countries have found hope and inspiration in this experiment; it has provided an example of the effectiveness of government intervention in achieving national strength in a short time. It has also provided a mechanism of doing so, the process of planning. The USSR instituted a series of five-year plans. The objectives and the various constraints are defined and from them policies are derived that hopefully lead to the achievement of the desired goals.   In theory, the planning process provided the national leaders of developing countries, who had formed new governments, with a modus operandi of transforming the colonized society. Many governments such as India, Pakistan, Sri Lanka, and Nigeria formulated mid- and long-term plans and developed an administrative setup for this purpose.(14)

Around this time, particularly after the Great Depression, another idea was gaining acceptance. This was the Keynesian idea that government policy can effectively improve the economic conditions in society. Prior to the publication of General Theory, the paradigms of classical economics found no useful economic role for the government. The economic system based on market mechanism was considered self-regulating. Self-regulation did involve booms and depressions. However, the severity of suffering in the depression was so widespread that there was need for an outside regulator. Many economists argue that Keynes saved the market system. If the self-regulation of the market system involved regular depressions of this kind, the political system would not allow the self-regulation mechanism. Keynes's ideas provided an important role for the government in managing investment and effective demand. Roosevelt's New Deal was an expression of Keynes's idea.

Also around this time, in the 1940s, national accounting methods were developed to measure the concept of GNP (gross national product). These measurements provided information on a number of other economic variables, such as national consumption, investment, savings, and so forth. The quantification of these macroeconomic variables provided operational significance to the Keynesian system and gave a further boost to these ideas. Around this time, macroeconomic theories of growth were also developed.(15) An interested government, thus, had all the intellectual and quantitative means to devise economic policies.

4 Conventional Development Strategy (CDS); Basic Propositions

BASIC PROPOSITIONS

As the developing countries were gaining independence, four influences were at work. There was a general acceptance of the idea that the government can, and should, influence economic policy, in a positive way. A number of developed countries were enacting legislation that required their governments to devise policies in order to achieve, and maintain, full employment. A Full Employment Act was passed in the United States in 1946. A few years later the English parliament adopted the Beveridge Plan, and similar measures were adopted in France, Germany, and the Netherlands.

The general population in the developing countries aspired to better material standards of living. National freedom movements stressed the idea that colonialism was exploitive and a direct cause of the poverty of the people. These people hoped, and even expected, that once independence was achieved and a national government formed, exploitation would cease, and total production would stay within the country and be fairly distributed. In the freedom struggle, they perceived not only an effort to get back their national pride and manhood (or womanhood), but also the possibility of an improvement in the quality of their living.

National leaders were interested in making their countries strong and in showing to the international community, particularly the colonial countries, that they could govern, and that their country was strong. They believed that national strength and improvements in standards of living are the same thing, and they felt that national strength had to be developed by energetic efforts and policies of development.

After the Second World War, the balance of power shifted heavily in favor of the United States and the United States exhibited little interest in creating and maintaining colonies of its own. On the contrary, it encouraged, even participated in, national struggles for freedom. Its interests, instead, were defined in terms of need for an

31

assured supply of raw materials and a large market for its industrial products.

All the above influences converged on the idea that national governments should actively engage in the process of development. The question was, and has been, how to define or determine development. Around this time, the prevalent notions of GNP became very useful. An increase in GNP per capita provided a handy and acceptable index of development. It suited the population at large because an increase in GNP per capita implied that everyone would be better off. It satisfied the national leaders because it treated the country as one unit without involving any political divisions within; it provided a useful and different role – it did not involve any fundamental changes in the status quo. All the changes implied were of the nature of add ons. Thus an explicit objective function would be defined in terms of raising the GNP per capita in a number of years.(1) Planning provided a useful tool, and a plan, or a series of plans, could be worked out. The government could, and would, formulate economic policy for development and introduce new economic and social institutions for this purpose. From this process emerged what we call the Conventional Development Strategy (CDS). It is called conventional because by now, after 30 years of operation, it has become the accepted convention. CDS is based, essentially, on two propositions that deal with growth and distribution.

Development can be defined as an increase in GNP per capita. Generally the relationship between GNP and welfare, particularly the narrow concept of economic welfare, is conceived to be positive. Strictly speaking, this relationship is more stringent since the implication is that increase in GNP has led to increases in economic welfare. Since GNP estimates are made for a country, an increase in GNP means that the economic welfare of the country has increased. But economic welfare of a country is not easy to define; it is not obvious, even conceptually. GNP per capita, on the other hand, relates the income and economic welfare of a person. This has, at least conceptually, clearer meanings. Since GNP per capita is a ratio, it involves two quantities: GNP and population. The development then follows from maximization of the rate of growth of GNP and minimization of the rate of growth of population.(2) Treating development in terms of GNP per capita suggests that changes in GNP and population are independent of each other.(3) It also leads to the problem of distribution, our second proposition.

Regarding distribution, the following proposition is implied.(4) Once production has taken place, it can always be distributed equally and easily.(5) A priori, this proposition seems convincingly obvious. If GNP has a particular value, a sum of money, the distribution is a simple act of division. Thus the real problem does not lie in the theory but in the mechanism of distribution.(6) It is argued that government fiscal policies such as selective taxes, subsidies, and transfers can take care of the problem of distribution. The interesting thing to note here is that this proposition completely ignores questions about the existing income distribution and the institutions that encourage or discourage a particular income distribution.

Given the above proposition that distribution will take care of itself, the problem of development translates into the growth problem: namely, how to achieve highest rates of growth of the GNP. Thus in CDS the maximization of the rate of growth of GNP is the key strategy.(7) Given this objective, further propositions follow. Since GNP is the aggregate of the quantities produced, more precisely values added, its rate of growth will be maximized if the goods produced are those that can be produced in large scale, and those that command high prices.(8)

CENTRALIZED MODES OF PRODUCTION

Large scale production has been argued on the basis of large fixed investment. Given fixed capital, as the scale of output increases, more and more of the fixed capital is used. The fixed cost per unit of output decreases until the scale of production reaches the capacity limit of the fixed equipment. Generally speaking, then, methods of large scale production are those that employ large quantities of fixed capital. Capital equipment is an embodiment of scientific and technological knowledge. Its design, operation, maintenance, full utilization, and replacement require various kinds and levels of specific skills. Scientists are needed to understand the scientific principles on the basis of which the capital equipment is designed. Specialized engineers are needed for specific tasks of construction, maintenance, operation, and so forth. Since the production is carried on on a large scale, supply and the specification of the material inputs becomes crucial. This generates the need for "purchasing specialists" and "materials engineers." Large production also involves a large number of skilled workers. This gives rise to specialists in labor management. When the output is produced, it has to be disposed of, along with various joint- and by-products such as waste. This means a host of other specialized tasks, such as marketing, waste disposal, utilization of by-products, storage, and transport. Since large capital involves large sums of money, this requires specialists in finance capital, banking, and investment. It is perhaps no accident that large-scale production has been, historically, resource, capital, energy, and skill intensive. The only thing it saves is labor. It also requires, and produces, urban areas. We call these methods of production centralized modes of production, because the control and decision-making in such production systems is highly centralized in a small group.

Large-scale production, undoubtedly, leads to increases in GNP. In fact, GNP estimation is biased in favor of these methods of large-scale production. The bias follows from the emphasis on specialization both in GNP estimation and large-scale production. However, from the point of view of the national leaders in developing countries, this development strategy had two additional advantages. The developed countries have been following the path of large production in the process of their own growth and development. As a result, these methods are prevalent in these countries. Even the USSR(9) (then a socialist country) had opted for the path of large-scale or centralized modes of production. Also, these methods had the stamp of "modernity." These methods have been called "productive" and "efficient," both terms based on output-labor ratio. The existence and growth of cities, of large visible projects, of specialized and skilled people, has an appearance of sophistication, as if it were coming of age. These additional reasons appealed to the "assimilative" responses of the national leaders. Accordingly, governments of developing countries have followed policies to encourage, establish, and foster centralized modes of production.

The logic of large-scale or centralized modes of production implies a number of social and economic policies. Large production units need large amounts of capital equipment. In economic language, capital involves savings. Savings are production not consumed.(10) Requirements of large-scale capital equipment raise two questions: how and where these savings come from and how these savings are translated into capital equipment. It has been maintained that savings come from high-level incomes. In other words, rich people save and poor consume, and thus the richer the people are, the higher the rate of savings. This argument assumes that actual savings are a linear function of potentials for savings. This proposition has been used to justify the existence, even the growth, of income inequalities and wealth disparities. To increase savings, not only should the policies accept income inequalities, but they should encourage greater inequalities by the provision of incentives to save and the establishment of opportunities to protect these savings. The second source of savings is the revenues of the government. Governments, all over the world, raise a major part of their revenues by taxing the majority; by sales, excise, custom taxes, and so forth. The savings policy here also ends up going against the poor and reducing their consumption. If enough savings are not realized, there is a constraint on the development of the strategy. In development literature this has been termed the "resource constraint" or "resource gap."(11) One way out is to obtain the savings from other nations via "foreign aid."(12)

The next question is how to transform these savings into capital equipment. In case a market exists for such equipment, this question involves the realization of "relevant money" from savings. (By relevant money we mean the money that this market will accept.) Thus, if the market is in the United States, the relevant money is U.S. dollars. Savings, then, have to be in a form which can be exported to a market from which U.S. dollars can be obtained. For the time being, putting aside the problems of exports, a host of questions arise.(13) What capital equipment should be obtained; and should the country produce machines to produce goods, or machines to produce machines, or machines to produce machines that will produce other machines which in their turn will produce consumer goods? From the very beginning this has been a controversy. The Western developed countries have moved from machines that produce goods to grandmother and great-grandmother machines. The USSR started from grandmother machines. The controversy still continues and there is no obvious or unique answer. There are a number of other questions regarding selection of technology which we will deal with in the next chapter.

To be operative, capital equipment requires other complementary factors – namely energy and skills. By and large, energy needed is "high quality" energy such as electricity where temperatures are raised to high degrees. Thus a complementary policy is needed that ensures a continuous and reliable supply of high quality energy. Skill requirements may be divided into two groups; skilled labor and specialized skills. Centralized modes of production require a pool of skilled labor which needs job training and training schools, on the one hand, and a market for labor. Specialized skills have a much longer gestation period. These are acquired at universities, colleges, and research institutions. The complementary policies, then, involve setting up engineering, science, and management schools and universities and the establishment of various research institutions and laboratories. Since universities take quite a long time to establish, development policy in CDS involves sending a large number of students to study in the colleges and universities of developed countries. A visit to an important university in any developed country provides evidence of the pursuit of this policy. And this policy also provides a basis for "brain drain."(14)

Centralized modes of production also require a reliable supply of materials, particularly of a specific variety. Since the quantities required are large, these have to be transported from different areas and final products have to be transported over long distances to markets. Both these considerations require a transport policy that provides infrastructure in terms of roads and railways, and that maintains the cost of transporting these goods at a relatively low rate. Such transport policy, naturally, results in large and continuous expenditures that have to be justified on other grounds.

Once centralized modes of production are created and reach a certain threshold level, they develop their economic and political constituency (vested interests by another name). Because of the centralization of economic power in a few hands, the constituency is able to wield large influence on various government policies; particularly those affecting its existence and growth.(15) Thus it generates its own dynamics.

HIGH-PRICED GOODS

Another implication of increasing the rate of growth of GNP is to produce those goods that have higher prices. This proposition involves a number of other propositions. The most important argument here relates to the process of monetization, or introduction and extension, of exchange relations in the economy. The effect of introducing exchange relations is to assign money values or prices to goods and services for which there have been quid pro quos but no prices.(16) The estimation of GNP is particularly biased in favor of money relations. For example, the services associated with the concept of a "housewife," where one person provides a number of different services and goods without any exchange for money, are not recorded. However, as soon as money is introduced into these relations (for example when the same services are provided in the form of a restaurant) these get recorded in GNP and the money prices of these services go up. The effect of the introduction of exchange relations is, certainly, to raise money prices.(17)

Money prices involve a market system. In any society the prices are determined by income distribution.(18) Generally speaking, in all developing countries the income distribution is highly unequal. Because of this unequal income distribution there are certain goods that get high prices. These are the "luxury goods" demanded by the persons at the upper end of the income distribution. Similarly, international prices are determined by international resource distribution. The distribution of resources in the world is also highly unequal. The developed countries control and consume a large part of world resources. In view of this, there are goods produced in developed countries which, in terms of the economy of developing countries, have high prices. These are the "imported goods" – imported by the developing countries from developed countries. Conceptually, the vector of "luxury goods" is different from the vector of "imported goods." However, there are a large number of common elements in these two vectors.

Even if we know that "luxury" and "imported" goods have relatively higher prices, the CDS proposition implies that production should be extended to these goods. The policies emanating from CDS encourage the production of luxury goods. Since the prices of luxury goods are high and since these can be obtained only by a small part of the population with large incomes and wealth, the policies imply the existence of income inequalities. They even encourage it. In addition to development goals in the manner of CDS, income and consumption inequalities are justified on other grounds as well. In the last section we referred to the arguments based on need, and capacity, for savings. Another argument is derived from the theory of incentives. It is maintained that the society can, and should, be mobilized by a plethora of economic incentives. The consumption of luxury and other high-priced goods provides the necessary incentives for work effort. The incentive argument is reduced to an ideology at the popular level; namely, those who work hard and produce more deserve to be paid more so that they can buy more.(19) The assumption is made that the larger the income difference, the higher the work effort. The probability of obtaining high incomes is ignored.

ISSUES OF IMPORTS AND EXPORTS

The existence of income inequalities and the CDS policies in favor of creating centralized modes of production and production of luxury goods involve the imports of a large number of different goods and services by the developing countries from the developed countries. These imports may be classified as consumer goods, capital goods, capital investment, technical know-how, skilled labor, persons with specialized skills, and so forth. The large variety of these imported goods and services raises a number of questions: what, where from, and how much to import. Also, do these imports satisfy incentive functions; are these imports inputs in the production process; and if so, are these essential inputs? If imports are essential to the production process, then development, or the rate of growth of production, will depend upon the level and availability of these imports. Such necessary imports create a "dependence" of the developing countries on developed countries. The level of imports, then, imposes a "constraint" on development. Other important aspects depend on whether these are one-time imports, or if they are needed every year. Will these imports increase or decrease as the scale of domestic production increases? Some of these are specific questions and can be answered only for given specific situations.

In the section on centralized modes of production we have suggested that imports are needed for large-scale production methods. A number of development theorists, particularly those in developing countries, have supported the proposition that imports are an essential input in the process of GNP growth and hence development.(20) They have also maintained that lack of these imports, or development, results in a relationship of dependency of the developing on developed countries. To remove this dependency, they have suggested a variant of CDS known as the "import substitution industrialization."(21) It involves the creation of domestic production capacities in all the goods and services which are considered essential and belong to the import vector. The basic idea is that once these industries are established, the need for imports of these goods will be reduced and eventually eliminated. A priori, this argument seems obvious, but it is not as simple as it looks. First, the setting up of production capacity, particularly on a large scale, of any goods is costly. The logic of setting up capacity is that the scale of production will be large enough to allow production at the minimum average fixed cost, and also the complementary factors will be available.(22) If, for some reason or another the capacity remains unused, it has two unpleasant effects. The average fixed cost of production is high, and some of the investible resources are frozen and are not available for productive uses. There are a number of reasons why capacity may not be fully used – lack of demand; lack of ancillary factors, facilities, and materials; technical change and obsolescence. Also, identification of the elements of the import vector into essential and nonessential is not an easy task, particularly in view of the interrelationships in production. This task is further complicated by the fact that development is a dynamic process. As a result, the elements of the import vector keep on changing. The setting up of the import substitution industries may itself create demand for other imported goods. It is possible that the import substitution policies may lead to more, instead of less, imports of goods, thus making the country more dependent.

Imports are possible if, and only if, the importing country has resources in the foreign currency. The CDS, therefore, suggests the following policies. Export as much as possible to developed countries, in order to earn the necessary foreign exchange. The developing countries have been exporting raw materials and/or some agricultural commodities such as cocoa, coffee, bananas, rubber, tea, and petroleum. These were the exports that were developed during the colonial period. The policy of encouraging exports, then, involves the extension of production in these commodities. Such exports face two problems" substitution of the natural by industrial or synthetic product as in the case of rubber, and fluctuations in the price of these commodities in the international markets. Fluctuations of price have been serious and thus this question has been raised many times at international forums and now forms an element of the New International Economic Order. In spite of these new problems, there are CDS variants based on the idea of export promotion. The "export-led theory" of growth derived from the theories of "gains from trade" is such a variant.(23) The theory argues that exports provide incomes via profits to the entrepreneur and the employment to the workers. It is assumed that these incomes can, and are, spent in the domestic market. Through the multiplier effect, these incomes generate a process of development. The implied assumption is that there are strong backward linkages. In case the economy had developed at some level via import substitution policies, the export promotion provides a different process. The exports are now industrial goods not raw materials and agricultural commodities. These goods compete with the industrial goods produced in the developed economies, and export promotion policies seek the opening of such markets. The New International Economic Order also contains elements favoring such policies. The export promotion and related policies at this stage of development constitute what is now being called the "Brazilian Model" of development.(24)

Another policy suggested by the CDS relates to the fact that if export earnings are not sufficient to cover the imports deemed essential, the developing country faces a "foreign exchange constraint" on its development. The only solution left is to seek "private foreign capital" and/or "foreign aid."(25) Private foreign investment has been institutionalized by such international bodies as the World Bank and multinational corporations. The policies that attract private foreign capital involve making the investment by the foreigners profitable and secure. Foreign aid, on the other hand, involves bilateral relations between a developed and a developing country. The policies that attract foreign aid deal with the interests of the individual donor country. Since the developing countries have faced shortages of foreign exchange, it is no accident that the United Nations Conference on Trade and Development has regularly sought a declaration by the developed countries to provide a small percentage (1 percent) of their GNP in aid to developed countries.

Variants of such policies have been, and are, prevalent in virtually all developing countries. Since the leverage of production comes from investment, the objectives of maximizing the rate of growth of GNP translates into policies for investment. Plans in all the countries have followed the practice of increasing the level of investment. In addition, an increasing proportion of this investment has been concentrated on providing overheads that facilitate industrial production. The investment in agriculture has also been directed to the capitalization of agricultural production.(26)

It will be noticed that the effect of CDS is to create the developing countries in the image of developed (and formerly colonized) countries. This is consistent with the "assimilation" responses in the colonial relationship.

5 Science and Technology Under CDS

ROLE OF SCIENCE AND TECHNOLOGY

We have argued in the last chapter than CDS places heavy emphasis on the production of goods and services on a large scale. There are two ways that goods and services can be produced in this way. One is production by the masses. In this form each person is working and producing individually at his/her own pace in an environment defined and determined by these people themselves; individually or jointly. What makes the production in large scale possible is the involvement of a very large number of people.(1) These production techniques require decentralized modes of production because the decisions to work and produce are made by workers themselves.(2)

Mass production is another way. It involves the division of the production process into small repetitive tasks that can be performed mechanically. The decisions to produce and divide production into various tasks are in the hands of a few people at the top of the hierarchy. Factory production and assembly lines are some of the examples of mass production techniques. In our terminology these techniques represent centralized modes of production and it is these forms of production that involve the applications of science and technology as it is commonly understood. We must emphasize the term "commonly understood." We do not mean, or imply, that decentralized modes of production do not involve application of science and technology. They do. However, the nature of these applications will be different in the two modes – "centralized" and "decentralized." CDS clearly implies centralized modes of production or techniques of mass production.

The moment one introduces the idea of the applications of science and technology a few questions arise. What is science and technology, and how are they related? How are science and technology involved in industrial production? Webster's New Collegiate Dictionary defines science by referring to these following four aspects: knowledge obtained by study and practice; any department of systematic knowledge; a branch of study concerned with observation and classification of facts and hypotheses; specific accumulated knowledge systematized and formulated with reference to the discovery of general truths or the operation of general laws, especially such knowledge when it relates to the physical world. This definition provides a philosophical concept of science. In practice, the observation and classification of facts involve a capacity for controlled experimentation, otherwise science remains purely theory and hypothesis. The advances in scientific knowledge follow not only from pure theory and hypotheses but also, more important, from greater capacity for controlled experimentation. It is in this capacity for controlled experimentation that technology comes into play. Another important dimension of science relates to objectivity. Any researcher using scientific method should be able to observe the same phenomena and obtain the same result. Objectivity implies, and involves, reproducibility. The experiment must be capable of being reproduced.

Webster's defines technology as industrial science, applied science, and systematic knowledge of the industrial arts. The interesting thing about this definition is that it relates technology and science insofar as it considers technology a form of science. This definition of technology is, obviously, contrary to the commonly held notion that technology is different and follows from science. This notion has been perpetuated by the practice of considering machines as technology and knowledge as science. This notion is simplistic, mistaken, and false.

It is difficult to distinguish between science and technology; both are highly interrelated and interdependent.(3) One does not follow the other, and even historically, technology has not followed science. In many instances it has preceded science. The steam engine provides an interesting example. It resulted from pure tinkering and made possible many advances in science. The advances in sciences today would be impossible without enormous technological preparations. Advances in, and even the understanding of, such fields as nuclear science, aerospace, heart surgery, and cell biology, depend in an essential way on the existence and availability of technology. Technology, on the other hand, is by definition an applied science. It contains all the elements of science and thus it is difficult to distinguish between science and technology in practice. In this book we treat science and technology as one.

Technology is also defined as industrial science. It has both dimensions of science and industry, particularly industrial production. Just as technology has been effective in the development of and advances in science, it has been an essential ingredient in the increase of the scale of mass production and in the introduction of "new" goods into production; both implied in the CDS. Thus science and technology play an important part in CDS.(4)

TWO CONCEPTS OF TECHNOLOGY

In social science and policy literature technology has been described variously, both at micro and macro levels and there is an extensive literature. One can discern clearly two different concepts of technology, a narrow concept in terms of "techniques of production," and a broader concept in terms of a "process." The latter concept includes the former.

The idea prevalent among economists has been that technology leads to a shift in the production function. In other words, it involves an increase in production of goods for given resources, or else it involves the production of the same amount of goods by a reduction in the factors of production.(5) This idea is based on the narrow concept of technology. It treats technology as a "technique of production" or as a machine, disregarding everything else such as social relations, infrastructure, and so forth. It implies that such techniques of production can be applied anywhere and has given rise to the idea of "transfer of technology."(6) Thus, if a developing country desires to produce goods in larger scale, the policy should be to follow the techniques of production that lead to increases in production for given resources. This concept has been in wide use and still finds a number of adherents.

However, this concept is obviously limited. It is even false, insofar as it assumes too many givens. It assumes that all the other ancillary factors are given. It gives technology a touch of manna, as if one can produce something from nothing, and there is now serious questioning about this concept. There is a large and inconclusive controversy among economists about whether technical change is embodied or disembodied.(7) Sociologists also find this concept wanting, if not misleading.(8)

There are two serious problems here involving the relationship between technology and production and omission of other factors. This concept implies a positive relationship between technology and productivity – the more sophisticated the technology, the higher the productivity. Yet this relationship is not obvious. For example, heart surgery and the atom bomb use some of the more sophisticated technology. Yet it is difficult to state that these are highly productive. The second difficulty goes against the fundamental law of economics, "there is no free lunch." This law is also considered a basic law of ecology.(9) By stating that technology can produce something out of nothing implies that there is a free lunch. However, even the ardent neoclassicists will not state the advantages of technology so categorically. In that case, this concept depends on the method of accounting, which resources are included and which are not. One recognizes some of the various problems with this concept as soon as one analyzes the cases of technology transfers.

The development of technology is a historical process. It acts and interacts with the society, in terms of products desired, resources available, criteria of efficiency, and mechanisms of choice. The second concept of technology is a broad and inclusive one; it takes into consideration all the linkages, forward as well as backward. The technique concept of technology assumes no linkages. The process concept inolves input techniques along with their management on the one hand, and products, sales, and distribution on the other. It is a dynamic concept with its own laws of motion, including the earlier concept of "techniques of production," or machines, or technical efficiency. In view of the process concept, technology integrates techniques of production and consumption with the society.(10)

For purposes of analysis in this book, this process concept is a more meaningful one. We have suggested CDS as a path; it is a series of logistical choices involving technology, educational institutions, management structures, transportation networks, fiscal and monetary policy, and the like. It determines the type of technology deemed appropriate for fulfilling its goals. In its turn, setting and maintaining this technology in place strengthens the apparent requirements for the whole array of CDS-related infrastructure. Thus technology, as a process concept, meshes with the CDS path.

Another advantage of this broader concept of technology is that it explains the various problems and difficulties encountered in the transfer of technologies. Technology transfer has been promoted by the concept of technology as a technique of production. However, when the techniques were transferred, it was found that these techniques were not as productive as expected. In addition, these have given rise to some unforeseen problems.(11) To salvage the situation, and meet some of these problems, a number of social and economic policy measures had to be adopted. Such measures certify that technology is not a technique but a process.

In view of the above reasons, in this book we use the process concept of technology.

CHOICE OF SCIENCE AND TECHNOLOGY

We have defined technology as a process. The questions now arise of what determines this process, and how technology is chosen. Here a number of factors are involved. On the demand side the questions are: who are the decision makers; what are their objective functions; what resources do they have; and what are the criteria for choice? There are similar questions on the supply side. Who are the suppliers? What technologies are available at a point of time? What determines what is available? In view of the fact that supply and demand for developed countries is different from that in developing countries, there will be differences in the choice of technologies by the different groups of countries. Figure 5.1 below highlights some of the issues involved.

Let us start from the demand side and from the point of view of developing countries. Theoretically, the demand of science and technology in a country can be derived from the "needs" of the society in the country. At any moment of time, any society has a multitude of needs. Some of them are basic, such as food, shelter, clothing, security from foreign attack. There are a large number of other needs and what these needs are depends upon who defines them. If we express all these needs by a vector, the elements of this vector are then dependent upon the social and economic institutions responsible for need identification.

Obviously the vector of the needs of a society is a large one. In any society, a reasonably large number of these elements is defined by

Fig. 5.1. Supply, Demand and Choice of Technology

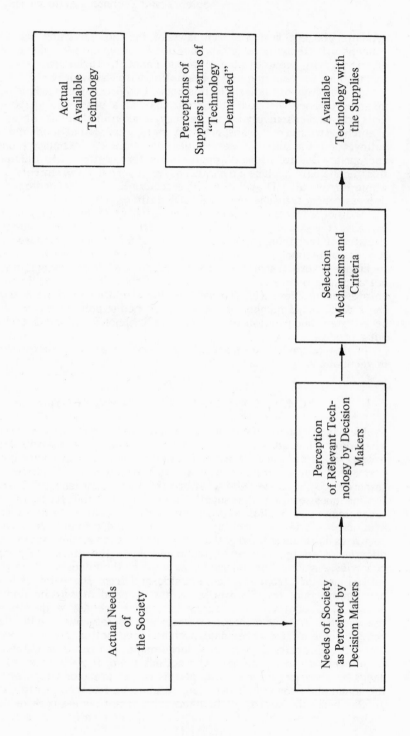

political leaders, senior bureaucrats, and upper-level business executives. If the income distribution is particularly unequal, some of these
elements are also defined by wealthy people.(12) It is particularly so in
the developing countries. This small group of people may be called the
decision-makers; they decide what is needed and what is not needed.
The number of decision-makers forms a very small fraction of the
population in developing countries, one in a thousand or perhaps less.
The issue is not that a large number of elements in the need vector is
defined by the decision-makers. The issue is what weights are ascribed
to the elements identified by the decision-makers. If sufficiently high
weights are ascribed to these elements, then virtually the whole vector
is defined by them.   Put in another way, the decision-makers have
certain perceptions about the needs of the society which may or may
not reflect the needs of the society.(13) How reflective these
perceptions are of actual needs depends upon a host of factors, mainly
economic and social institutions and the distribution of economic and
political power. As we have suggested in Chapter 3, the decision-makers
in the developing countries have been particularly influenced by the
"assimilation" characteristics of colonialism. As a result, their need
identification is affected by what is happening in the developed
countries. Accordingly, they may give greater weight to certain goods –
used in the developed countries – in the needs of their countries. In view
of their acceptance of the CDS path and the formulation of growth
targets, they have in fact translated the needs of the society in terms
of the growth rate of the GNP. Thus it is the perceptions of the
decision-makers about the needs of the society that determines the
objective function of the society, or the country. This objective may be,
and many times is, made up of a number of varied, sometimes even
contradictory, objectives.

   The choice of technology depends essentially on the objective
functions of the society. They define the type, nature, and character of
the technological process. For example, if the objective function is
national prestige in hosting an international conference then large
meeting halls and necessary facilities will be built. This was done in
Kenya to hold UNCTAD IV.(14) The objective function, however, sets
the outer limits to the choice of particular techniques. Actual
technological choice depends upon another factor also – the perception
of the decision-maker about what technologies are relevant to meet the
needs of the society and what are consistent with available resources.
The perceptions about relevant technologies are, in their turn, determined by what may be called the "science and technology establishment." It consists of a subset of the decision-makers who have some
expertise in science and technology. In developing countries this
establishment is very small indeed; it may be approximated, at the
limit, by the total number of scientists and engineers. (It will be noticed
from Table 2.2 that this is a very small group indeed.) The smallness of
this establishment places heavy strains on the choice and introduction
of technology and technological decisions thus become an extension in
this establishment. In addition, if modern technology by itself is one of
the perceived needs of the society, the development and growth of this

establishment becomes an objective of policy. Scientific laboratories, science and engineering colleges, and sending students to study in developed countries are the institutions that eventually extend this establishment.

As the supply side is concerned, there are at any moment of time a number of technologies that exist in the world. Some of these technologies have been in operation in the past; they still exist even though they are not in operation. Some of these technologies are in actual operation in some part of the world. Some are small technologies with little or no linkages; others are large with enormous linkages in various sectors. Some technologies are in the experimental stage; others have passed through this stage but are waiting to be in operation. All of these technologies define the set of total available technologies which is being continuously extended by R&D and educational institutions involved in science and technologies. Most R&D is being done in developed countries (15) and a large number of science and technology institutions are also in the developed countries.

From among all these technologies, only a small number form a large part of the actual supply of technologies. These are the "technologies on the shelf," and they also relate to some of the components of the "technological process" as we have defined it. In view of our concept of technologies, these are not complete except when they are small technologies with no linkages. But which of the available technologies are selected for operation? Let us call the class of people who make this decision of choice suppliers. They select those technologies which they feel will satisfy the demand, but they may not be the technologies in actual demand. In the market system, these suppliers will prepare these technologies for eventual sale to those who demand. If their perceptions are wrong, they can end up holding those technologies that are not demanded. However, if they can also influence demand, then they can pass on the technologies they have prepared. This may be true of multinational corporations or government contractors.(16)

We notice that demand for technology comes, inter alia, from the developing countries. This demand is interpreted by the science and technology establishment which is generally biased in favor of large technologies with enormous linkages. The supply, on the other hand, is concentrated in the developed countries. The suppliers of technology in developed countries, particularly large technologies with enormous linkages, are made up largely of multinational corporations. Even though there is a separate group of buyers and sellers, there is no well-defined market in technology. There are a number of reasons for this deficiency. There is no standardized large technology; every large technology is a particular one. In view of the specificity of technology, there is also a lack of knowledge and information on such technologies. Only some of the components of such technology are well-defined. The science and technology establishment, in developing countries, is also handicapped because of lack of sufficient resources to appraise and evaluate particular technologies. The evaluation of large technologies is a costly affair. Another reason for the lack of a well-defined market is that suppliers of technology are not uninterested bystanders in the market. They are interested in trying to influence the buyers.(17) In view of the specificity of the technology, the influencing is possible, and it can also be hidden. The market is also ill-defined because there are no objective criteria of choice such as efficiency, price, and productivity. All these criteria are relative. Efficiency can be defined in the manner of the first or second law of thermodynamics.(18) But the conclusions are completely different. Productivity depends essentially on which factors are considered and which are excluded.(19) As regards prices, there are so many different prices that one can justify any particular choice.(20)

In view of the above, choice of a particular technology depends upon such a large number of factors that it is difficult to predict which particular technology will be chosen. However, the science and technology establishment and the suppliers do define the limits within which these choices are made. It may also be added that the choice of technology is also a process so that it affects the establishment and suppliers and is affected by the choices already made.

NATURE OF EXISTING SCIENCE AND TECHNOLOGY IN DEVELOPING COUNTRIES

Many developing countries have chosen industrial technology as the optimal means to growth. Such a choice has followed from the imperatives of CDS. This technology can be used to fulfill the demand that may exist for highly priced consumer durables, to produce manufactured goods for export to world markets, and to produce goods locally that can substitute for imported products.

A major problem is that the technology linked to this output has generally been developed by innovative processes and hard work in countries whose factor endowments differ in intensity from their characteristic of most Third World countries. That is, the technology of mass production and large economies of scale require relatively high capital/worker ratios, high use of energy, and thoroughly trained personnel to deal with the complexity of mechanization and support services and associated R&D facilities. These requirements have been met in developed countries which have large high-income markets, a large supply of scientists and engineers, and a propensity to increase productivity by substituting (at least until recent years) equipment based on relatively cheap capital costs for relatively expensive labor.

These conditions are more or less absent in many developing countries which feature a lack of mass consumption markets, large pools of unskilled labor, unsophisticated R&D backup, and a scarcity of capital needed to finance equipment investment and corollary energy generation systems. Yet it is technology engineered for the different circumstances of the developed countries which has been pervasively transferred by them to developing areas under the CDS in the hope of reducing underdevelopment. Why has this mismatch taken place?

First, there is a general perception in many countries that the "best" technology, in the sense of most appropriate to their factor endowments, is the newest, most mechanized technology, and that a country lacking such technology cannot compete in world markets. Nonmodern technology, in this perspective, connotes being mired in economic stagnation and technological backwardness, and in a wider sense, not being part of the forward movement of the "advanced" sectors of civilization.

Second, factor price distortions are common in developing countries to the extent that the relative prices of labor and capital do not reflect the pricing system that ought to exist based objectively on their comparative scarcity.(21) Investment decisions are biased by a comprehensive, and in part unintentional, pattern of government policy decisions toward large industrial technologies. These decisions include capital subsidies, preferential tariffs or none at all on the import of capital equipment, tax holidays for large foreign enterprises, privileged access to credit markets, government training programs and foreign educational programs oriented toward high technology, commitments to high visibility, prestige projects, and the like. Looked at from the other side of the coin, locally based, small scale entrepreneurs using labor intensive technologies are likely to have a difficult time in establishing themselves out of proportion to their objective economic condition. In short, investment decisions take place in an atmosphere wholly conducive to the import of readily available technology from abroad, with serious consequences for employment rates, income distribution, and prices of goods (which may be high in protected markets) in the recipient nations.

Third, the CDS and investment decisions place a premium on attracting the resources – capital, personnel, and technology – of the multinational corporations (MNCs), whose orientation toward promoting their own capital intensive technology rather than adapting it to local conditions further promotes developing country dependence on technology mismatched with their factor endowments.(22) The absence of incentives favoring adaptive engineering, the lack of a large cadre of native technicians who can critically analyze local engineering needs, and the centralization of most R&D carried out by MNCs within their home countries and laboratories located at their headquarters all feed into the importation of off-the-shelf technology.

Fourth, the conditions under which technology is transferred make it difficult for developing countries to select other than modern industrial machinery.(23) These conditions include MNC "packages" which wrap technology together with associated capital, management, and marketing services; MNC contractual requirements limiting host country policies on importing competitive products, exporting products manufactured with MNC equipment, and diffusion of imported technology to other domestic forums; and tying of purchases of needed spare parts, raw materials, and other products to the MNC. This bundle of links between developing countries and MNCs results fundamentally from asymmetrical bargaining conditions and heavily influences the range and scope of technological choices open to the former.

Finally, these choices are further limited once the MNC is ensconced within the host country through its ability to avoid or drive out price competition with small local entrepreneurs engaged in producing equivalent products. Oligopolistic markets result from the policy biases noted above in favor of the MNCs and imported technology, and from MNC promotion of brand name familiarity. Again, a result of driving out indigenous business is to increase the technological dependence of developing countries.

In reaction to these conditions, many third world governments have sought relationships with MNCs not involving direct investment or integrated packages in order to regain control over types of technology desired. Thus, lease or sale of patents, joint ventures, and other institutional reforms may be attempted. More broadly, pursuant to the call for a new international economic order, there have been demands for a "Code of Conduct for the Transfer of Technology" that would refer to the restrictions of recipient country flexibility on the use of imported technology noted above.

But a more fundamental issue arises, or rather remains. Whether or not developing countries maintain control over their technological choices in the context of relationships with MNCs and foreign aid donors, one must still inquire about the appropriateness of the technology itself to the basic human needs of the societies involved.

6 Experience of CDS

DEVELOPED COUNTRIES: FACETS OF OVERDEVELOPMENT

It is the Western world which pioneered the application of CDS. It epitomizes the goals of maximizing material consumption, high growth rates, and the presumed benefits of conventional technology. The United States and other developed countries have successfully pursued CDS over a long period of time. It is this success that defines them as "developed countries."(1) These countries have achieved levels of material production and consumption so high that consumption has percolated down to the majority of the population in these countries. For large segments of the population, the problem of poverty has certainly been solved. In these countries, there is a large middle class whose material standards of living are quite high.

However, the development in these countries has not been smooth or balanced. Just as the United States has been particularly successful, it is also the United States that dramatically illustrates facets of overdevelopment as a result of CDS. It has pursued a CDS path which Kenneth Boulding has called the "cowboy economy."(2) It has been based on ever-expanding linear flows of consumption and production, and it has meant consuming a disproportionate share of the world's resources, as well as the inefficiency with which these resources have been gathered and transformed into consumer and industrial products. Overdevelopment arises from at least three interrelated sources.

1) CDS assumed unlimited supply of resources. The economies became resource-dependent and they have lost their flexibility to solve some of the problems. On the contrary, this resource-dependence has created its own problems.

2) Large scale production units have been promoted on the basis of economies and efficiences of scale. There are now growing diseconomies and inefficiencies.

50

3) There has been an assumption of a positive relationship between growth and human happiness, but GNP growth has created environmental problems.

We discuss these sources of overdevelopment below.

Resources

The consequences of wasteful resource exploitation and of the use of CDS technologies are well-illustrated in the energy sector. The era of cheap energy in the United States has had a pervasive influence on modes of industrialization, transportation, agriculture, and lifestyles in general. This has become all too apparent in light of a five-fold increase in the price of Middle Eastern petroleum since 1973.(3) This, in turn, triggered increases in the costs of gasoline and petroleum-related products, including food. United States agriculture is heavily dependent on oil for fertilizer, pesticides, packaging, harvesting, and transport, and the importation of oil is approaching half its domestic consumption. A majority of these imports are from the Middle East. The global production of oil and gas is expected to reach its peak in the 1980s, thus posing threats to the availability of this resource. The availability of nuclear power to provide an increasing share of United States energy is coming under serious question because of continuing problems of waste disposal, regulatory delays, and capital costs. Environmental problems are effecting potential increases in the mining of abundant coal.

The consequences of these trends are obvious: inflation, a plummeting American dollar, an oil-induced gap in the balance of payments, dependency on possibly unstable sources of oil, and threats to the functioning of industrial and urban infrastructures developed when energy costs were low and there were few inducements to efficient energy use.

Energy is not the only area in which formerly abundant, or easily available, resources have become more scarce under the brunt of exponential demand, increased problems of accessibility, and unwillingness of local populations to bear the cost of resource exploitation. These conditions characterize a number of important minerals needed by the United States and other industrialized nations. Even clean air and water and valuable ecosystems — such as grazing land, coastal zones, and offshore fisheries — face limits upon their use as productive resources, or dumping grounds for residuals of industrial society. Thus there is a range of materials which, in the middle-term future, will not be available to developed countries as a whole at any reasonable price.(4)

Apart from the matter of resource availability, historically there has been an apparent correlation between a rise in the rate of energy generation and its use, on the one hand, and growth in employment opportunities and the GNP, on the other. This presumption still holds a powerful grip on policymakers and the public generally. However, the concept that energy and employment rates, or energy and material well-being as a whole, have directly paralleled each other and will continue to do so if enough effort is put into continued energy production has been seriously challenged by a number of critics of the cowboy economy. These critics contend that the logical result of increasing the use of energy is to reduce employment, particularly in the industrial sector where energy intensive mechanization has traditionally been used to lower costs by reducing the need for high-wage human labor.(5) Thus, major energy-producing and using industries consume one-third of the energy in the United States, but directly provide only 10 percent of the jobs. Some analysts have calculated that small changes in transportation and consumer goods would increase employment and reduce energy use.(6)

Energy, of course, has increased labor productivity, i.e., output per worker. However, it has reduced the total number employed. This effect has been masked in the past by an increase in employment stemming from greater consumer demand from a rising population. Manufacturers needed to add more workers, while those displaced might find jobs elsewhere. The reason for present rates of unemployment in the United States and other industrialized countries cannot be simply explained, or easily understood.(7) An important factor seems to be that advances in automation, and accompanying energy rates, are displacing more workers than can be absorbed elsewhere in a faltering economy.

The issue of capital investment raises similar concerns. Enormous sums of money are required to finance large-scale centralized sources of energy generation and energy intensive manufacturing processes. These create relatively small numbers of jobs. Lovins estimates that investment in proposed energy systems would total three-fourths of all private investment capital available in the United States through 1986, approximately one trillion dollars.(8) Even if it could be made available, other social uses are competing for this capital, such as medical care and rehabilitation of urban infrastructures. Interest rates are raised when large sums of money are tied up in long-term investments. The large deficit payments incurred by the United States to pay for imported oil have also contributed to capital shortages and unemployment.

Scale

It has been argued that efficiencies are gained by large-scale production and manufacturing activities. Many scholars have denied any inevitable impact of increasing scale of endeavor on improved efficiency of operation.(9) Indeed, there may be a reverse relationship in effect. Large-scale enterprises carry their own costs, economic and noneconomic, in the form of increased bureaucratic red tape, less inclination to take risks, decreased technological innovativeness, and hierarchical workplace relations between employers and employees. All these factors may lower whatever advantages lie in economies of scale in terms of lower production costs and, therefore, increased profitability in a narrow sense. Diseconomies of scale for big firms and high technology must also be extended to include greater vulnerability to breakdown, with more serious second-order consequences in the event. There is also less flexibility in the face of changing consumer demands. These diseconomies are becoming all the more effective in view of greater consumer awareness and resulting pressures for increased corporate accountability.

The concept of efficiency must be particularly broadened to encompass thermodynamic or end-use criteria, by which nonrenewable resources would be treated as precious capital stock; losses in the generation and transmission of energy would be minimized; and polluting "externalities" stemming from industrial processes would be reintegrated into the production cycle or eliminated.(10) Put in other terms, it is a strange sort of efficiency which sees environmental degradation, increased unemployment, and higher rates of usage of scarce resources as signs of economic success.

Civilization Malaise and Ecology

Even though material consumption has increased enormously, a certain level of poverty still persists in United States urban areas, and communities have been in continuous decay. Enhancement of the viability of urban communities and income maintenance for their poor inhabitants are, thus, highly salient issues for United States development. Low-income people are spending 53 percent of their total income on food, compared to a national average of 20 percent, and 15 percent of their income on energy. Poor people spend more of their limited budgets on food and energy than do those better off, and often obtain lower quality products for their money.(11) Traditional welfare is becoming increasingly costly and it does nothing to relieve the economic and psychological burden of continued dependence on government support. At the same time middle-class consumers, correctly or not, perceive welfare spending as a major contributor to their tax rates. This has become a source of continuous tension in the society.

Those who are able to afford advanced levels of material consumption are realizing that this material progress is not costless.(12) To obtain these incomes, they find that the work is dull, drab, and dreary. In view of the uncertainty of the incomes, there is a lot of tension and the work is truly alienating.(13) On the other hand, material consumption is impinging on time and making it still more scarce. The scarcity of time is changing the very culture of the society.(14) The community and, therefore the human relationship is in perpetual decay.(15) Crime is on the increase, spreading from cities to suburbs, and there is an increasing atmosphere of fear. Life in old age is, at best, dismal. As a result of these factors, increased levels of material consumption are not satisfying the human spirit. There is a growing skepticism whether any increases in material consumption can even satisfy the human spirit. This recognition has been termed "civilization malaise."(16)

Industrial production in the United States has strained the ecological balance. The environment movement, highlighted by the earth day in

1970, brought this fact into focus. Commoner has argued, persuasively, that technology is one of the major causes of environmental degradation.(17) Since then there is a growing awareness about the various harmful impacts of technology on environment. Thus, many days every year people living in Los Angeles, and other cities, are advised not to breathe hard, because of the pollutants in the air. It has been stated that drinking water, in many United States towns and cities, may be injurious to health. Some of the popular foods may have harmful chemical reactions.(18) The number of chemicals introduced into the environment is quite large and one hears about the bad effects of these chemicals. PBB, PCB, and Love Canal are the most recent examples.(19) There is thus a general sense of uneasiness.

DEVELOPING COUNTRIES: PERSISTENCE OF UNDERDEVELOPMENT AND EMERGENCE OF DUAL SOCIETIES

By now the CDS has been in operation for 20 to 30 years in a large number of countries. In this period we have witnessed two United Nation's sponsored decades of development. There is, thus, ample experience on the application of CDS and it is time for taking stock. In the areas that are directly related to it, CDS has been quite successful. The GNP per capita has grown and these rates of growth have been higher than in the previous decades under colonialism. There has been an enormous increase in the introduction of new technologically sophisticated goods in the economy. Thus, one can now go to virtually any country in the world and find an airport, automobiles, luxury hotels, luxury housing, large prestigious buildings, and so forth. These would have been unknown only a few decades back. These successes have also created new problems, and have accentuated old ones. By and large, one can establish the following stylized pattern in countries where CDS has been applied.

In sociological terms, this pattern of development has created a dual society composed of two separate societies. One is a very small fraction of the population, made up of the rich, elite, urbanized, and powerful. The other is the large majority of poor, exploited, rural, and powerless people.(26) We shall call these societies R and P respectively.

A. 1) Industrial production has grown tremendously.(20)
   2) There has been an overall growth in energy consumption involving a switch from solar to fossil fuels, and coal to oil within fossil fuels.(21)
   3) Road transportation has grown much faster than rail transport.
   4) There has been a growth in the size and number of large cities and a movement toward urbanization.

B.. 1) Agricultural production has kept pace with population.(22)
   2) Use of fertilizers has grown tremendously.

C. 1) The number of poor people, as well as the proportion of poor among the total population, has increased.(23)
   2) Unemployed and underemployed, both in number and as a proportion of the total working force, have increased.(24)

D. 1) Production structure has become highly skewed in favor of production of goods for the rich.
   2) Consumption inequalities have widened.(25)

E. Population, on the whole, has increased and continues to increase.

DYNAMICS OF A DUAL SOCIETY

In economic theory, dualism has been considered one of the processes of development.(27) Typically, a society is divided into two sub-societies, modern and traditional. These are equivalent to our R and P. The dual societies R and P are distinguished in terms of the production function and the market for labor. The production function of the modern sector is neoclassical involving high substitution between capital and labor. The labor market is assumed not only to exist but to be well defined so that wages are determined by the marginal productivity theory. The wage rate offered in R is assumed to be much higher than that in the P society. In the P society production function is land-determined and there is no well-defined wage market. Wages are determined by average product which is lower than the marginal product in the R sector. The whole purpose of the exercise is to study the "desired" process of transference of labor from P to R and the introduction of money relations in P society.(28)

Policies derived from these theories have been applied. Unfortunately, they did not work. Part of the explanation is that the theory has not been sufficiently developed to incorporate some of the major distortions.(29) More seriously, critics have argued that this theory completely missed the fact and practice of imperialism and neocolonialism.(30)

The dual society in developing countries has developed its own dynamics with twin objectives – to ensure the existence of, and to sharpen the divisions between the two societies, R and P. There is a serious question whether these two objectives are in conflict.(31) Our interest here is in the dynamic process itself. This process is based on four characteristics of the R society, richness or economic power, political power, elitism, and urbanness.

R possesses most of the economic power of the total society, in terms of income generated and wealth. Since the population size in R is comparatively small – it forms only a miniscule fraction of the total society, less than .01 percent – the per capita income in R is rather high. It compares favorably with average per capita incomes in the Western developed countries. The pattern of consumption in R society, thus, is compatible with consumption in the Western developed countries. In a market oriented economy, this economic power defines the pattern of demand which in its turn determines the pattern of production, investment, imports, and exports. It is this economic power that explains a ready, even black, market for imported goods from the Western developed countries. It provides the justification for an import substitution and export promotion policy.

Political power is also held in the R society. Even though the potential of political power lies in the P society, the mechanisms of this power are held in R.(32) This power can be, and is, used to help maintain economic power in the R society. The mechanisms here are government incomes policy via taxes and government expenditures. The taxes and subsidies tilt the factor prices in favor of the pattern of production desired by the R society. The elitist character of the R society favors government decision-making towards the production of prestigious goods and monuments. Thus the production and development of "high technology" becomes an issue of highest priority. Such activities bestow economic advantages to the R society in the form of high salaries, large profits to contractors, importers, and producers. The government expenditures thus generate, and accentuate further, the pattern of demand, production, investment, imports, and exports suited to the economic power in the R society.

Elitism has two effects. It generates a pattern of consumption that further divides the two societies, P and R. The members of the R society consume goods and services not available to the P society. Such consumption favors the import of goods from abroad, preservation of goods out of season, and production of goods involving high energy and based on high technology. Part of the reasons for elitism lie in the "assimilative" dimension of colonialism.(33) The members of the R society want to be accepted by their erstwhile colonizers, and distinguish themselves from other colonized people.

The second effect of elitism is that it favors expenditures in production of prestigious goods and projects. Thus large projects involving cities, dams, and high technology are desired for their own sake even when they make little economic sense in terms of cost-benefit ratios. The whole developing world is full of such large expensive white elephants, and even the neoclassical economists are forced to point out their inappropriateness. Elitism particularly favors capital intensive projects.

Since the R society is urbanized, this characteristic helps in the location of large projects and production units in the urban areas. In the developing world, the rate of growth of cities has been the highest even though these form small pockets in otherwise large rural areas. It needs to be pointed out that in urban areas there are a large number of urban poor. By urbanized, we refer to a very small fraction of the urban population who constitute the urban elite.

The four characteristics satisfy the conditions for existence of the R society; they also sharpen divisions between R and P societies. Thus production is undertaken in urban areas so that urbanization remains strong. Prestigious production and consumption further enforce elitism and keep the two societies apart. The pattern of consumption, production, investment, and imports ensures that mechanisms of economic and political power remain in the R society. However, the dynamics of dualism also involves some other factors such as property relations, an international economic and political order, and high technology. Our interest here is limited to the technological factor alone.

Our question is about the role technology plays in the dynamics of dualism. Given the characteristics of the R society, it chooses (perhaps naturally) that particular technology which suits and promotes its dynamic processes. It so happens that the technology developed in the Western developed countries is especially suitable to the preservation and growth of the R society.(34) This technology produces high energy intensive products, and is itself capital, resource, skill and energy intensive. It is also known as high or hard technology.(35) This technology satisfies the economic, elitist, and political needs of the R society. It serves economic interests insofar as it provides high income to managers and capital-owing members of the R society. It serves political needs in view of the centralized control that hard technologies require. All such technologies are large and prestigious. They are also urban based. Even if these are located in the rural areas, in the first place, they soon create an urban environment because of the nature and scale of production.

THE WIDENING GAP BETWEEN DEVELOPED AND DEVELOPING COUNTRIES

As we have noticed, the developing countries have made special efforts to accelerate their growth rates. They were hoping that they will also be able to catch up with the developed countries. If the catching up is more time consuming, the hope has been that the distance between the developed and developing countries will be reduced. To do so, major efforts have been made at the international level. Part of the objectives of UN development decades has been a reduction in this gap. UNCTAD was created in 1964, and since then it has held four conferences, an international conference every four years. UNCTAD has made attempts to seek increased "aid" from the developed for developing countries; to obtain commodity agreements that stabilize the prices and earnings of the exports of the developing countries; and to encourage policies by the developed countries to open up their markets to the industrial products from the developing countries by reducing quotas and tariffs. However, the gap has continuously widened. The developing countries, in the group of 77, have expressed their frustration and determination in the form of seeking and obtaining a UN general assembly resolution in favor of the establishment of a New International Economic Order. This gap has also affected what is now called North-South dialogue.

It is not obvious why this gap has widened. There are various possible explanations. The most popular explanation, particularly in the developing countries, is that of imperialism and neocolonialism. In other words, it is the natural result of the extension of capitalism at an international scale.(36) Currently, the Marxian school is in a state of division. The Old Left accepted the USSR as a socialist state. Recently it has been declared that the USSR system represents state capitalism, and the government has also been described as a "social imperialist." China was considered a socialist state by the New Left. After Mao's death, there is now increasing skepticism about the possibility that the Chinese State has moved back to a bourgeois state. It is thus difficult to recognize all the nuances in the theories of imperialism and neocolonialism.

Another hypothesis is based on the divergent rates of growth of "international demonstration effect" and "technical change."(37) The argument is based on a dynamic process. Let us start with technical change. Imperatives of the Galbraithian industrial society maintain that the corporations in the developed countries keep their rate of technical change by producing "new" technical goods. Because of the international demonstration effect, these new goods are transferred, at a fast rate, into the patterns of demand of the R society in the developing countries. This changes the composition of demand in the developing countries and the effect of this change in demand is to reduce the demand for substitute goods being produced domestically. In view of the highly skewed income distribution, and lack of a middle class, the producers in the developing countries cannot sell the substitute goods. They are left with unused capacity, high costs, and a declining market. On the other hand, these new goods, or the technology to produce these new goods, has to be imported. By the time the developing countries are able to catch up, the industrial structure in the developed countries starts the next round, and thus the gap widens.

7 CDS Evaluation

CONCEPTUALIZATION: DEVELOPMENT AND GNP

As we have noted, implementation of CDS has tended to accompany increases in poverty, unemployment, and income inequality in recent years. Inevitably, economists and others have had to look at the meaning of development. The meaning could once be taken for granted, or it appeared so obvious as to warrant little thought, but that is no longer the case. The problem lies in the automatic equation of development with maximization of GNP, and of the latter with the amelioration of poverty. In fact, GNP is an inadequate measure of development because there is an asymmetrical relationship between poverty and GNP.

It is true that if GNP is measured on the basis of a number of poor people living not as a cohesive social unit, but as individuals, their GNP per capita will be low. However, the converse does not follow. If the GNP per capita of a group is low, it does not imply that the group is composed of poor people or people who suffer from poverty. This asymmetry arises from the methods of estimating GNP, which is based on the exchange relations of goods and services entering the market. For something to be included in the GNP, there must be attached to it a monetary value or price. Otherwise, it does not (literally) count.

However, there is no reason that all goods produced by a society must be so valued. Not everything is offered for a quantifiable price, and thus GNP figures for a country may underestimate the totality of products in circulation. More to the point, GNP is incapable of evaluating a highly cohesive social group which is predominately nonexchange oriented, sharing goods on the basis of barter or other arrangements. Such a group, which does not have profit maximization as a primary goal, need not be impoverished. Of course, this implies a value system that puts a premium on cooperation, mutual aid, and collective responsibilities.

In this vein, it should be noted that the introduction of exchange

60

relations can destroy viable social groups whose relations are not defined by exchange. Industrialization has this effect. Since GNP is based on and biased toward exchange relations, industrialization is automatically associated with increases in GNP. Thus a paradox results, and the introduction of exchange relations between persons is not necessarily an improvement, but may be a deterioration in the conditions of the poor even while GNP is rising. There is now ample evidence on this proposition.(1)

In addition, GNP does not note the type of goods produced. The production structure of a dual society produces luxury goods such as refrigerators, air conditioners, automobiles, and expensive houses, but poor people need food, clothing, and shelter. This is not to say that poor people do not wish luxury goods. They do, but their first priority, however, is to satisfy their basic needs. However, they lack the means (money) to translate their needs into market demands. That is, "goods" produced may not be responsive to at least the survival needs of the poor.

Approached from the other direction, it would be hard to argue that citizens of the affluent countries are well off or happy in direct proportion to increases in their GNP. Past a certain point, the accumulation of material goods becomes a meaningless preoccupation which does not lead to lasting personal satisfaction and may correlate with deteriorating personal and environmental health.

What, then, is development? We will address this issue in the following chapter in the context of a development strategy oriented toward self-reliance, employment, and the provision of basic human needs.

ASSUMPTIONS AND OVERSIGHTS

One interesting thing about this strategy is that it gives hardly any consideration to either the needs of the people, especially poor people, or their employment. Given the CDS predilection for large, urban-based development projects whose benefits will somehow find their way to the poor, it is not accidental that references to people and employment are indirect.(2) The employment issue is especially crucial since, under CDS conditions, most people must earn the wherewithal with which to purchase needed goods. Gainful employment becomes a primary link to ensuring one's ability to get along. Yet, as Chapter 4 has explained, CDS emphasizes centralized modes of production and the turning out of high priced luxury goods because of their contribution to the GNP. But such production processes are capital and skill intensive – they are not intended to provide for mass employment. Thus many individuals are simply shoved out of participation in the exchange system and must seek marginal, substandard living arrangements. The theory of benefits and job opportunities trickling down from the modern sector to the rest of the population does not work, but must be addressed directly. For this reason any alternative strategy must place priority on employment stimulation through basic goods production.

The CDS also assumes that distribution of whatever goods are produced will take care of itself, or can be easily taken care of. This also does not stand analysis. A production system implies a distribution system. Other things remaining equal, which the CDS is intended to ensure, it is the distribution system in place which will determine who gets what, when, and how. As we have seen, this system in most developing (and developed) countries is riddled with income and political inequalities which work to distribute CDS products inequitably. Clearly, once goods-not-needed by the poor are produced, there is no government or market mechanism that can distribute these to the poor, nor is the system geared to producing goods the poor do need. Production of luxury goods, at least in the initial stages of development, implies a denial of goods in general to the poor if problems of distribution are not tackled directly. In short, under CDS, the rich get richer and the rest scurry along.

By fixing no goal other than an ever increasing expansion of GNP, the CDS must assume the continuing availability, even at rising rates of demand, of the necessary resources, the energy by which to obtain and use them, and the capacity of the environment to absorb the resultant pollutants. As noted above and in Chapter 1, all these factors are being queried within developed countries, and so is the issue of whether economic growth requires concomitant energy growth in the first place. This is not to take issue against developing countries receiving their fair share of resource needs, and more than that, to make up for historical inequities. It is to argue in favor of environmentally responsible development paths – what the UN calls "ecodevelopment" – that benefit the poor. A repetition of affluent countries' consumption patterns is available only to the rich in developing nations, and as we have argued, cannot be sustained in any case.

The CDS has nothing to say explicitly about questions of justice or equity in relation to development, and this is not surprising since it supports the continuance of the dual society. Equity considerations are not only moral issues. Even in "practical" economic terms, a society committed to egalitarian and participatory goals is more likely to attain continuing balanced growth rates than otherwise. The issue is not whether equity is compatible with growth, but whether growth can be achieved without simultaneous steps toward equity. In the absence of such steps, widening of the gap between rich and poor within developing countries may lead to a social unrest which hampers productivity. Growth can be aided by encouraging participation by the people in productive processes which they control. Ironically, narrow-minded concern for material growth, to the exclusion of all other factors, works against itself.

GNP assumes the effectiveness of prevailing institutions and political systems. This is one advantage of equating development with GNP rates, but the advantage is short term. If inequitable conditions impact on a development program's chances for success, it makes sense to pay attention to the institutional framework which perpetuates social injustice. Strategies which ignore this framework are not likely to make much dent on the needs of the poor. This is precisely the charge that developed nations have raised in response to Third World demands for the New International Economic Order (NIEO). Why should dramatic shifts in resource allocations and aid take place from developed to developing countries if institutional patterns in the latter only guarantee that most of the benefits will accrue to those already well off? It could be argued that developed nations have taken this stance as a ploy to avoid serious negotiations on the NIEO and that developing countries as a whole have legitimate demands to make regardless of their internal situations. Therefore, alternatives to the CDS must deal with development processes as well as program content.

CDS assumes that off-the-shelf technology provides the best means for expanding GNP; indeed, acquisition of the latest technology becomes an end in itself. We have examined the biases built into the conceptualization of development and the policies designed to make this so. But if technology imported or produced according to the CDS does not, in fact, serve to alleviate mass poverty and if it is based on resources whose continued availability is in question, then notions of what is the "best" technology for developing countries must be challenged.   In this respect, CDS is characterized by technological tunnel vision.   Both the types of technologies and the means of acquiring them are much wider, as we shall see, than is presumed under CDS. There is more than one "shelf." While one must travel outside the neighborhood to locate the traditional shelves, alternatives can be located close to home.

Beyond technology, the entire thrust of CDS points in the direction of heavy doses of aid and other transfers from developed to developing countries. CDS, then, assumes that developed countries are able and willing to help. Without them, there is no viable CDS. But this is a shaky assumption indeed since official development assistance from developed countries remains, for the most part, parsimonious.   In the United States, as in other countries, the foreign aid budget is always easy to attack and has no automatic constituency to defend it. Concerns of the Third World do not receive sustained attention in the media. After several years of haggling around the NIEO, developed countries have reacted negatively to such proposals as automatic resource transfers and debt moratoria, or reacted half-heartedly to proposals for commodity stabilization funds and other more modest demands. Within developed countries, opposition has arisen from labor and other quarters over the issue of relaxing tariffs that restrict imports of manufactured goods from developing countries. We do not excuse developed countries from contributing fairly to equitable development plans, and we ask what is to be done if they do not, and if political pressure brought to bear by developing countries is unavailable or unavailing. We believe the answer lies in the direction of development strategies that emphasize national and regional self-reliance.

Finally, some of the economic relationships which the CDS presumes have not worked out in practice are highlighted here. Income inequalities do not lead to increased savings. Actual savings are not proportional to, or a linear function of, potential savings. This is illustrated in the case of the rich who indulge in conspicuous consumption as a result of the international demonstration effect – keeping up with the Joneses in the affluent nations. Thus, savings from their high incomes do not materialize. Instead, it is the poor who end up saving involuntarily through inflation and indirect taxes.

Incentive systems implied in the justification of income inequalities do not function. There is no direct relationship between hard work and high incomes. The latter arise, generally, from elements of monopoly and "rent." Even when hard work and income are related, the probability of obtaining high incomes is low and insignificant as an incentive. Backward linkages do not exist. Export promotion schemes end up generating incomes that are siphoned out of the domestic economy either via imports or through profits to importers/manufacturers abroad. Even when the incomes in the export sectors are available for domestic purposes, the levels are so low that they are not able to set up a dynamic multiplier process. The time profile for such a process extends over centuries.

Import substitution policies end up establishing high cost, large-scale production units, thereby bringing into question the validity of the decreasing cost assumption of the large unit. This follows from a number of factors such as the limited size of the market and rate of change of technology.

DEPENDENCY

Even if the CDS were not riddled with fallacious assumptions, there would remain an important problem. Once operationalized, it works to heighten or create certain types of dependency relationships both within developing countries and between developed and developing countries. Such relationships are at the heart of the asymmetrical distribution of the costs and benefits of global interdependence referred to in Chapter 1. It is our contention that such inequities are inherent to CDS and can only be altered through adoption of alternative strategies.

Within developing countries, the CDS gives rise to several sets of dependencieswhich may be summarized as follows. Rural areas, where the bulk of the Third World population lives, become dependent on the cities. Industrial projects are located in the latter, as are the financial, technical, and educational infrastructures which undergird them. It is in the city that relatively high paying jobs – and perhaps any jobs at all – may be found, as well as the accoutrements of modern life. The city, then, drains the countryside of those who might otherwise prefer to remain there, but who receive no support – credit, technical assistance, marketing, services – for doing so. But there are not enough jobs for all who seek them, especially on the basis of unskilled labor, in the cities. The results are vast urban slums abutting pockets of affluence, a recipe for ecological and social disaster.

The uneducated become dependent on the specialists. Chapter 4 traced this phenomenon directly to the type of technology and mode of production generally required by the CDS, and Chapter 5 indicated how such technology is mismatched to the factor endowments of developing countries. In dependency terms, the majority of the population must rely on the abilities of those who are able to maintain and obtain advance technologies. Under these circumstances, traditional or newer technology better adapted to environmental and cultural conditions of developing countries become forgotten or ignored. The great majority who are unschooled feel like, and are perceived as, drones, unable to contribute to the life of the nation. Education is not the same as wisdom, but under the CDS there is little premium placed on attempting to learn from, and seeking the participation of, those who remain alienated from the modern sector.

Women become dependent, or more dependent, on men. There is, of course, nothing in CDS that theoretically excludes women from its productive relations. But this tends to be the case, as in devloped countries. This has been noted with particular regard to agriculture. In many developing nations women have a primary role in rural agriculture, but modernization programs are geared to men. It is also women who bear the brunt of continuing rural life as best they can when men who can depart for the city.

Between developing and developed countries analogous dependencies exist. Psychologically, as indicated in Chapter 3, assimilationist tendencies originating in the colonial era continue even after formal independence has come to the ex-colonies. The rich in poor countries wish to surround themselves with the same material ambiance available to the citizens of rich countries, and so they adopt a development strategy that appears to have worked for the West. Indigenous values and ways of life that might provide an alternative framework for defining development are scorned in the rush to become modern. Ironically, at the same time, growing numbers in the overdeveloped countries are seeking out perspectives from which to judge the value of an existence committed to material accumulation.

Technologically, developing countries obtain the latest products and productive tools from developed countries and the multinational corporations headquartered in them. But the process does not stop there. A technology brings with it a set of relationships; it must be maintained, serviced and updated. Such peripheral services may only be available abroad; so are the necessary experts, consulting firms, and spare parts. To get around this drain on foreign exchange, a country may send its students abroad for the requisite training; they may or may not return home to apply it. When a developing country imports an advanced technology, it imports a way of life whose upkeep may be more expensive than bargained for.

Economically, the CDS draws developing countries into world markets dominated by the developed nations, and into the economic orbits of the multinationals. Developing countries must turn to these sources for investment and concessionary aid, while working to open up markets in developed countries for their manufactured goods and to obtain more stable prices for their raw material exports.   The only dependence that the rich countries sometimes feel is on such materials. If they have an inelastic demand in the production structure of the rich countries, this dependence can be heavy, as has been demonstrated with

OPEC oil. Otherwise, the costs of dependence fall on the developing world – literally in the form of huge debts which many poor countries owe to developed ones. A more subtle effect is alterations in the internal market structure of developing countries when they orient products to the taste of developed country buyers (and their equivalents at home). A common example is the use of agricultural land for export products intended to earn foreign exchange instead of for local food production. As we have seen, technology transfer packages sold by multinationals may also tie the hands of developing country planners and entrepreneurs, inhibiting their full use of available technology and their search for alternatives.

The causes of these various forms of dependency are inextricably linked to the CDS, not so much in terms of any one set of factors such as technology or financing, but as that strategy is conceived as a whole. This point may be clarified by a review of how the CDS works in various stages of a country's development.

In the initial stage of CDS, the developing country does not produce commodities desired by the rich countries. In view of the relations established during colonialism, all it exports are raw materials or particular crops. The available technology is rather limited. Thus, the effort of the government to accelerate development forces it to import from the industrialized countries. Since there are no exports to pay for these imports, the capacity to develop is limited by this gap in export earnings and import needs. This is the trade gap in the two-gap theory of development.(3) The only way out is to secure loans or aid from the rich countries. It is because of this gap that in the first meeting of the UN Conference on Trade and Development in Geneva there was an emphasis on aid from the rich countries, targeted at 1 percent of national income.(4)

In the middle stage, the country has developed some of its export potential, e.g., mechanized raw material extraction, and organized export crafts and crops. It has increased its capacity to export and earn foreign exchange. Since a large part of the exports are concentrated on a few raw materials and/or crops, changes in their price can make a large difference in the country's export earnings. We have seen that developing countries recognize the severity of this problem and have made minimization of price fluctuations and associated uncertainties an important element of the New International Economic Order. Given the potential of export earnings, the level of development is affected by the goals of the government. If the government desires to import more than export earnings, it goes back to the situation in the earlier phase, i.e., back to foreign exchange deficits. On the other hand, if the foreign exchange earnings are quite large as in the case of OPEC, imports are not the problem; the problem is investment of these earnings.

In the third stage, the country is not only producing for export, but also has extended production to substitute for earlier type imports. Much depends on this import substitution strategy. If it is successful, the proportion of imports decreases and eventually the foreign exchange gap disappears. But many times this strategy simply foists on the country a costly production sector which requires more, rather than less, imports. At this stage, another problem crops up. The products of the import substitution sector may not be fully sold in the country's domestic market. These products must then seek outlets in the markets of the rich countries.

In conclusion, we might ask what is wrong with dependency relations. Nothing, leaving morality aside, is wrong if one is on the receiving end of the benefits; a great deal, if one is not. Most people in the world are in the latter position. The end result of several decades of CDS is relatively modest improvements, and even some sliding backwards, in ameliorating Third World poverty, and a decided increase in friction between rich and poor countries, as well as between social classes in the latter.

Even on narrow economic grounds – on its own terms – the CDS does not work. But the costs of such interdependence go beyond this. It leaves developing countries out of touch with themselves and the unique cultural and technological contributions they might make toward a genuinely interdependent world. The CDS mortgages their future, leaving the bulk of Third World peoples dependent on institutions and forces, within their countries and abroad, that are unreachable and unaccountable.

III

Alternative Development Strategies (ADS)

8 Alternative Development Strategies (ADS): Basic Issues

NEED FOR ADS

It is now being generally recognized that CDS is not a solution to the problem of development in the developing countries. Even its protagonists contend that the time needed for the "trickle down" to reach the poor is long. On the other hand, in view of the varieties of goods produced and made available, as well as the accentuation of consumption inequalities, CDS promotes a rise in the expectations of people. The gulf between expectations and realizations provides a continuous source of political tension. A democratic political system finds this time too long to be acceptable to the people.(1) If continued, this tension creates political instability. There is, thus, a pressure to develop authoritarian forms of government. Critics of CDS, on the other hand, wonder if the "trickle down" will ever reach the poor. In their view, the success of CDS requires large income inequalities and hence the existence, and its persistence, of a sizeable section of the poor people in the most affluent countries. Furthermore, CDS involves high resource intensities. The United States alone, with 6 percent of the world population, consumes 40 percent of the world's resources, particularly metals and petroleum whose known reserves are not great. If the whole developing world were to have half the resource intensity of the US, the developing world would need 300 percent of the world's resources annually.(2) Critics question the availability of resources in these magnitudes, and if CDS can even have its full potential. There is now a growing awareness that resources of the earth are finite, particularly in terms of the ecosystems and ecosphere. There is a general consensus that it may be impossible for the developing countries to ever achieve the per capita material production standards that were prevalent only recntly in some of the rich developed countries.(3)

Even if one grants that the CDS can have its full potential, there is another equally important consideration. In the industrialized countries where the growth in material standards has been high, it is being

71

recognized that these high material standards have not brought greater satisfaction, happiness, and social harmony. There is, thus, a serious doubt about the relationship between material consumption and welfare. The object of development, in the final analysis, is welfare and not material consumption. If there are negative elements in this relationship, then the objective of material consumption and production, hence GNP maximization, becomes misconceived.(4) The GNP concept is under examination. There is a general consensus among economists that it is wanting, but there is no general consensus about an alternative to it. Samuelson has proposed the concept of Net Economic Welfare.(5) Amartya Sen has suggested a measure of development that takes into consideration income inequality and life expectancy.(6) The Overseas Development Council has suggested, and worked out, a Physical Quality of Life Index. This is calculated as an unweighted average of the indices of infant mortality, life expectancy, and literacy.(7) Interestingly, it does not include GNP at all. The debate on GNP is still not over. However, it does make the CDS, so dependent upon GNP, somewhat irrelevant. Skepticism about GNP has the effect of making one look into the composition of GNP rather than concentrating on the aggregate index.

CDS has not been able to reduce the growing rate of polarization of society in developing countries into two separate worlds of elite and poor. Many observers believe that CDS is far from neutral. Instead CDS has accentuated this polarization, and as it reaches threshold levels, it is becoming a major source of political instability, if not possible upheaval. It has resulted in brutalizing authoritarian dictatorial regimes, increasing repression, and denial of basic human rights. Argentina, Brazil, Chile, and Iran have become police states where torture and murder of prisoners have become a way of life. For two years, between 1975 and 1977, India also went through an experience as a police state. Police states are not only inhuman and indecent, they are inherently instable, and their instabilities have international consequences.(8) The oppressed people find that the only possible redress involves violence, and so the violence of the state causes a reaction of violence from the people. In this defensive violence, the people seek help from neighboring and other countries which leads to an escalation of violence at a regional and international level. There is a growing need, and recognition, that internal oppression and tension in these countries has to be reduced, if not completely eliminated. This implies that CDS has to be revised, if not replaced.

Today the very concept of development is under critical examination. The question is being asked: What is development? There is a developing consensus that development means the "development of a human being." CDS has also referred to the development of a human being but has emphasized only the material dimension of human development. There is no doubt that human development does require material good to satisfy some of the very basic physiological needs of a human being. However, human development is not based entirely on material goods and consumption. As Maslow has pointed out, human development involves an equal, if not larger, number of nonmaterial needs, such as security, belongingness, love, respect, self-esteem, and self-actualization.(10) It is thus essential that a meaningful development strategy should be able to provide an important place for activities and institutions that encourage such nonmaterial elements.

All of these factors suggest a strong need for an alternative development strategy (ADS). There is some thinking going on along these lines. However, concepts about ADS are still in their formative stage. The strategy is flourishing under various names – Appropriate Development, Basic Needs,(11) Gandhian Development,(12) Rural Based Development, Self-Reliant Development,(13) Small is Beautiful,(14) and so forth. Virtually all national governments and international agencies are involved in giving some of the elements of such a strategy shape and form. The ideas on such a strategy are still in ferment. There is, thus, little consensus on all the various aspects of such a strategy. There is, however, a lot of consensus on the facts on which ADS is based; the objectives of ADS; and the basic elements that constitute ADS.

THE FACTS ON WHICH ADS IS BASED

There is a general acceptance about the existence of the following facts in the developing countries.

1) A sufficiently large part of the population in these countries is poor. Poverty is defined narrowly. For example, in India a person is defined "poor" if his income is so low that it cannot purchase food from which he can obtain 2,000 calories a day (2,000 calories are considered, by the World Health Organization, necessary for an adult to perform normal body functions).(15) This is, however, too narrow a definition; here half the population is defined poor. However, if one adds a few additional, equally necessary consumption goods, the magnitude and proportion of the poor increases.(16) There are no easy estimates of the number of poor people. However, the ILO estimated 67 percent of the population to be seriously poor in 1972; approximately 1,210 million.(17) These people cannot satisfy their basic needs and they are exposed to natural and social disasters.

2) A cause of this poverty is unemployment and underemployment which is also widely spread. Underemployment is difficult to estimate and its estimates are likely to be underestimated. ILO estimates that in 1975, 283 million or 40 percent of the working force in developing countries was either unemployed or underemployed.(18) The proportions vary from country to country, the range is from 30 to 50 percent.

3) A very large part of the population in poor countries live in villages and rural areas. This is particularly true of countries

in the Fourth World. The percentage of population living in rural areas sometimes approaches 95 percent. Generally, the range is between 80 and 90 percent of the population. Since a large part of the population is also poor, the majority of the poor live in rural areas. This does not suggest that the problem of urban poor is less serious. Even in the urban areas, the majority are poor. Proportionally, a larger percentage of population is poor in urban areas than in rural areas.

4) There are not enough resources in the world to enable every country to achieve the United States levels of industrialization. Constraints are particularly evident in terms of environment and energy.

5) There are poor people who are either under or unemployed even in the First World developed countries.

6) There are enough resources, even in the developing countries, and certainly in the world at large, to provide all the poor people with at least a basic minimum standard of consumption consistent with human needs for food, health, and shelter. The greatest resource is the people themselves. There are estimates that suggest that the resources consumed by the United States alone can provide basic human needs to all the poor people of the world.(19)

7) There have been, and still are, large cultural diversities within and between various countries and regions. These cultural practices have in the past provided mechanisms and institutions for the production and distribution of goods for the satisfaction of basic needs. But these cultural institutions are in decay.

8) Even though people themselves are an important resource, people have become dependent upon solutions to their own problems which are presented by outsiders. They have not been taking the initiative to solve their own problems of unemployment, underemployment, poverty, and exposure to natural disasters. They are not involved; and they exhibit a lack of self-reliance.(20)

9) The natural environment in the developing countries has been mutilated so that various parts of the country and region are subject to natural disasters. It has been argued that population increases in India and Pakistan have made these countries more susceptible to natural disasters such as cyclones, heavy rains, and drought. The drought of 1974 exposed 400 million people in Asia and Africa to hunger.

10) There are widespread income, wealth, and consumption

inequalities within social groups and regions. These inequalities have in turn accentuated inequalities in other spheres, such as sex, age groups, occupations, and ethnics.

OBJECTIVES OF ADS

The objective of development, and of ADS, is the development of all human beings. However "development of a human being" needs to be distinguished from the "development of human potential." Development of human potential has been, and still is, considered as the attempts, achievements, and performance of unusually large tasks, such as building castles and temples, going to the moon, and so forth. The achievement of such tasks do bear testimony to human potential. However, many times such tasks have been achieved by coercion of large numbers of human beings. In this process, the development of human beings has been reduced. An example will make the distinction clear. The Taj Mahal, built in the sixteenth century, is still considered one of the wonders of the world. The story goes that it was built by 22,000 masons and workers working over 20 years. When the Taj Mahal was built, their hands were chopped. The Taj Mahal, no doubt, is a shining example of human potential, but the 22,000 who built it under coercion of the king virtually lost their humanity. They could not satisfy their human needs and so their own development was reduced. In the past, and in many developing countries today, coercion has been an essential ingredient in the development of human potential. Coercion, by definition, reduces the human development of persons coerced. This is not saying that the development of all human beings will eliminate all development of human potential. The development of human potential will still continue, and take different forms. Only those large tasks will be achieved in which people, whose human development has not yet been reduced or thwarted, will willingly cooperate.

The objective of ADS is the development of all human beings. The most essential element of such a process is the satisfaction of basic human needs. Basic human needs may be divided into material needs and nonmaterial needs, and we will analyze the objectives of ADS in terms of these needs. Material needs follow the physiological needs of human beings. To be able to live, human beings need food, shelter, and clothing. The objective of ADS, thus, is to satisfy physiological needs of all the people, particularly those whose needs have not been satisfied. These are the poor people and, as we have noticed, they form a majority of the population in the poor countries. The objective of ADS, then, is improvement in the welfare of the poor people. In other words, the objectives of ADS are defined by the welfare function of the poor. ILO has defined the objectives of ADS in terms of satisfaction of the "basic needs."(21) This objective relates to the important fact of widespread poverty.

Basic material needs are recurring. They have continued in the past and will continue in the future. Their satisfaction involves the use of material resources which also have to be of a recurring nature, such as solar energy. This objective of ADS, then, translates into the objective of recurring resource development. One of the most important recurring resources is human labor and ingenuity. The objective, therefore, is to involve this resource through work opportunities. This deals with the issue of employment and unemployment. At the same time, it can satisfy the nonmaterial need of self-esteem. It is, however, important how these work opportunities are created and organized.(22)

The recurring resource development involves, by and large, the development of renewable resources, and it emphasizes minimal use of exhaustible resources. It places greater value on the maintenance and improvements in stocks instead of their depreciation.(23) Such development results in improvements in the ecosystem since renewable sources are the products of a well-functioning ecosystem. In other terms, the objective of basic material needs translates into consistency with, and enhancement of, the environment.

The nonmaterial needs are sense of security, belonging, love, respect, self-actualization, and self-esteem. A sense of security and belonging comes from the existence of a community and a society. The objective of ADS, then, relates to the preservation of community and society.   Income, wealth, and consumption inequalities disrupt a community and society. These objectives involve the reduction and elimination of such inequalities, and address one of the important facts listed above. Some economists have articulated "distribution of income" as one of the objectives of development. The redistribution objectives range from an increase in the share of the poor from the increase in GNP to distribution of income levels and assets.(24) The preservation and integration of the community and society also promote self-reliance.   Some development theorists have advanced "self-reliance" as one of the objectives of development.(25)

Love and respect follow from the nature of a family which in its turn is an essential feature of a culture. The objectives of ADS, thus, deal with the value dimensions in a culture. It involves enhancing the valuable elements in a culture, promoting its dynamism, and stopping its decay. Such objectives also encourage self-reliance and are related to the objectives of preservation and integration of community and society.

Self-esteem and self-actualization lie in the realm of development of a human being. No human being can develop without self-esteem. One of the fundamental sources of violence in colonialism arises from the destruction of self-esteem. Self-esteem is fundamental if people have to participate, actively and cooperatively, in the development process. The objective of ADS, then, is to enhance and promote self-esteem and self-actualization. This means that it is not only the ends that are important; the means, or the processes, are equally so.

ELEMENTS IN ADS

Such different authors and protagonists of ADS emphasize different objectives, there is no one particular strategy or one given set of steps in ADS. Policies and mechanisms vary, but there is a consensus on the basic elements in the ADS.

The most important element deals with the production of material goods that meet the physiological, or basic, needs of all the people.(26) These are goods such as food, shelter, clothing, drinking water, and security. There is no one definition of basic needs. These differ from country to country, region to region, and even community to community, and they are defined by the people themselves and not be central and international agencies. Basically, this element deals with the issue of the composition of GNP. In view of the large number of poor people and the objective of satisfying the basic needs of all of them, ADS policies must lead to economic growth, i.e., growth in GNP.(27) However, this increase must come from the increase in the production of goods for basic needs, thereby changing the composition of GNP. In view of the change in the composition of GNP, an increase in GNP is not a sufficient indicator of the results of ADS. An increase in GNP can arise, and has arisen, from CDS policies. For appraisal of ADS one must look both at level and composition of GNP; both are equally important. A shift in the composition of GNP in favor of basic-needs goods involves, ipso facto, a redistribution in favor of the poor.

In terms of GNP, the end result of ADS is to increase the level of GNP and shift its composition in favor of basic-needs goods. The end results are important. However, as we have suggested earlier, in ADS means are equally important. It is, thus, equally important how this change comes about. ADS implies that such change in production takes place by means of decentralized modes of production, i.e., production by masses — to use a phrase from Gandhi. There are two obvious reasons in favor of decentralized modes of production. These deal, head on, with the fact and problem of unemployment and underemployment; and they lead to involvement by the people. The issue of unemployment and underemployment is a tricky one, and there is a serious debate on this issue. Many have arued that the thing that has gone wrong with CDS is the lack of employment creation. They are thus defining ADS in terms of generation of employment. Various projects are suggested in order to create jobs and employment, such as public works programs, road building, dams, and so forth. In other words, the arguments are being made, and are being seriously considered, that production processes should be made labor intensive. Looking at the problem of unemployment in terms of job creation, and labor intensity of production, is to ask the wrong question. Such a question is misconceived, and it elicits wrong answers. ADS does not imply the creation of jobs by some central authority. It means the provision of work opportunities for the people who are unemployed or underemployed. The assumption is that people are capable of employing themselves in case opportunities exist. The provision of work opportunities, however, does involve important social and economic changes. The creation of jobs does not deal with the second reason, people's involvement. However, provision of work opportunities also leads to active involvement by the people. Work opportunities promote initiatives by people in defining their own needs and their own work to meet these needs. In the process, one's nonmaterial need of self-esteem is also satisfied. Such initiatives involve participation by the people in decision-making, and encourage local solutions which in turn furthers local self-reliance. The problems of under and unemployment is solved and the income, consumption, and assets inequalities are reduced.

When the people define their own solutions through their own work to satisfy their needs, they derive these solutions from their own cultural traditions and institutions since these are the things they know best. This is a dynamic process which energizes culture which in turn becomes alive instead of decaying. Instead of one, there are numerous solutions varying with localities and cultures. The living and varied cultures provide a richness to the human being and to the society at large. Living cultures are excellent sources to satisfy such nonmaterial human needs as love, respect, and self-actualization.

The solutions people seek are local. Thus, ADS has a large element of production and distribution in local areas. The reason the solutions are local is because people find it easy to seek local solutions, and local solutions are also the most efficient solutions because these match resources to needs. The efficiency here is in terms of the second law of thermodynamics. Local solutions also have the advantages of improving the local environment and enhancing of the ecosystem. Since people live and work there, their interest in improvements is nothing short of self-interest. Such activities also earn one the respect of neighbors. Local solutions provide not only local initiative but also a sense of belonging and security; they promote what we have called "development of a human being."

Local solutions and cultural diversity do not imply the nonexistence of a nation state. The two concepts, and forms, are quite consistent. Localities are integrated into larger geographical units.

COMPARISON BETWEEN CDS AND ADS

There is a fundamental difference between CDS and ADS. It can be best articulated in terms of our analysis of the dual society. To recapitulate, we argued that some developing countries are composed of two societies, R and P. R is made up of a miniscule proportion of the population; it is elitist, rich, politically powerful, and urbanized. By comparison, P embraces virtually all the population; it contains the rural, poor, and powerless. CDS is defined by, and promotes, the objectives and welfare of the R society. It divides the country by breeding and propagating dualism. ADS, on the other hand, is determined by, and fulfills, the needs and welfare of the P society. It integrates the country into one society by reducing, and eventually eliminating, dualism. Some of these differences are explained below.

CDS has been rationalized on the grounds that poor people cannot, and do not, save; and that large-scale production – in factories and on farms – is more efficient. In view of these two assumptions, there is little reason for the bulk of poor people to participate in the

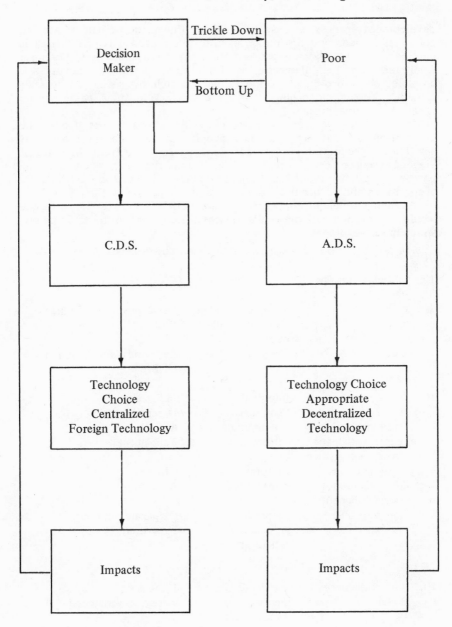

Fig. 8.1. A Schemetic Diagram Comparing CDS and ADS

development process because the bulk of the population in developing countries is poor and the major part of their productive activities is carried on in small-scale production. They have small land holdings, small farms, and small businesses. Their participation, whenever it is possible, therefore involves inefficiency which by definition reduces growth. People, accordingly, are a problem and their participation leads to inefficiencies.

In ADS this logic and the implied assumptions do not apply. The assumptions instead are that poor people can, and do save; and that small scale production is more efficient, particularly from the point of view of resource cost and the long run. This is the basic meaning of the phrase "small is beautiful." There is now evidence on both these propositions and accordingly, ADS seeks participation by the people.

At higher levels of concern there are still larger differences, as one would expect. Some of these differences are outlined in the table below, which is self-explanatory.

Table 8.1. A Comparison between the Characteristics of CDS and ADS*

General Characteristics	CDS	ADS
I. Objective	Maximum GNP per Capita or Welfare of R	Development of a Human Being or Welfare of P
Indicator	Level of GNP	Level and Composition of GNP
II. Technology	Imported	Indigenous
Modes of production	Centralized	Decentralized
Local institutions	Unimportant	Crucial
People participation in decision-making	Unnecessary	Fundamental
Local Solutions	Uniform	Diverse
Social change for people's benefit	Unnecessary	Necessary
Role of people vs. experts	People are the problem, experts are the solution	People are solutions, Experts are Advisors
III. Role of theoretical model	Standard theory is fundamental,** answers come from theory	There is no standard theory. Experimental

*We are thankful to Ted Owens for suggesting this table.

**Models such as Harrod-Domar, Libenstein, Mahalonobis are important. There are no well-defined models in ADS.

9 ADS:
Some Variants

The need for ADS is becoming more and more urgent. The ideas on ADS, on the other hand, are still in the initial stages of development, and there is no one ADS (nor could there be). There are, instead, various forms of ADS depending upon the emphasis on different objectives. Accordingly, we present below some of the variants of ADS.

IS WORLD BANK STRATEGY A VARIANT OF CDS OR ADS?

In 1974 the World Bank and the Institute of Development Studies of the Univeristy of Sussex in England completed a study that was published under the title Redistribution with Growth.(1) This study has been proposed as a variant of ADS. It outlines the basic philosophy, and hence the source of policy, of the World Bank in the sphere of development in general, and reduction of poverty in particular.

The World Bank variant of ADS concentrates on growth, as in CDS. However, it takes into specific consideration the existence of the poor. The poor are defined in terms of low-income recipients. In the rural areas, they are identified as landless labor, self-employed craftsmen, artisans, and those engaged in village services. In urban areas, these are the unskilled, unemployed, and self-employed engaged in low-productivity areas. The problem of development, then, is couched in terms of redistribution in favor of the low-income recipients.

Four approaches are suggested.

1) Maximization of GNP growth through raising savings and allocating resources more efficiently with benefits to all groups in society;

2) Redirecting investment to poverty groups in the form of education, access to credit, and public facilities;

81

3) Redistribution of income (or consumption) to poverty groups through the fiscal system or through direct allocation of consumer goods; and

4) Transfer of existing assets to poverty groups.

The basic emphasis is on a shift in the pattern and structure of income growth and asset ownership over time. The suggested policy instruments are "investment transfers" resulting from growth. The transfers are to be brought about by fiscal, credit, and other policies, including a reorientation of foreign aid.

The World Bank variant is "demand-defined." It depends, in a fundamental way, on the working of the markets. There is a philosophical contradiction in the World Bank variant. In the structuring of the income transfer it depends upon government intervention, recognizing thereby that the working of the market system will not ensure redistribution in favor of low-income groups. Yet, given income transfers, it assumes that the markets will work perfectly to produce the goods to meet the needs of the poor. The historical experience seems to go against both these assumptions. The fiscal and monetary policies have been particularly, and singularly, ineffective in redistributing resources from the rich to the poor; and markets have failed, generally, to deliver goods to the poor, even in years of plentiful production.

A serious question is: Does the World Bank variant contain large elements of ADS? Or is it an attempt to put old wine in new bottles? In this respect, Teresa Hayter's study is very relevant.(2) Her arguments are cogent and well substantiated and her conclusion that the World Bank is part of the problem (of the growth in poverty and unemployment), rather than part of the solution, is both serious and significant. Also Congressman Clarence Long's contention that the World Bank staff, generally, lacks the capacity to appreciate, comprehend, and understand the problems of poverty and unemployment is appropriate.

To examine further the World Bank strategy, we note its basic argument. The poor people, because they are poor, cannot enter the market and thus cannot generate demand. One way to provide incomes to them is to offer them employment. This is possible if investable funds are spent on projects that generate employment for them. In a poor country these investable funds are limited. Investing these funds to create employment only may sacrifice growth; growth is an increase in the GNP in the manner of CDS. It may be economic to invest these funds to maximize increases in growth, without regard, of course, to the composition of GNP. However, once growth is maximized, the fruits of this growth are transferred by income distribution mechanisms to the poor. They will demand the goods they need. The market will produce these goods. Hopefully, it will start a process that will eventually remove poverty. This is a nutshell summary of the "redistribution with growth."

A priori, this argument seems reasonable, even convincing. However, closer examination reveals a number of flaws. Let us start at the end.

The poor have received incomes by income redistribution. They spend these incomes on the goods they need. If the supply of these goods is not increased, their expenditure will result in bidding the prices of these goods up and enriching the suppliers of these goods. Will the suppliers invest these extra profits to increase the supply of these goods? There are two issues here. The first deals with the fact that the suppliers belong, generally speaking, to the richer section of the society. Their profits will increase the demand for goods in the luxury sector. As a result, part of the redistributed incomes have returned to the rich.

A major study provides evidence on this thesis. Sinha et al. (1978) have analyzed various schemes of income distribution in India and their impact on the poor. Their model is based on a 76-sector input-output table. Distinction is made for bottom, middle, and top incomes for rural and urban areas separately. In their own words, "the harsh conclusion thus emerges that, within the process of income generation and distribution as we have modelled it, the rural and urban poor are both relatively disadvantaged in terms of their evolving share in the income of their respective areas. But as a consequence of the urban bias in the pattern of incomes generated, the relative disadvantage suffered by the rural poor is particularly pronounced. Even with very substantial national growth, and with no allowance for population increase, the income available to the rural poor will be quite inadequate on the given distributional pattern, to provide for their basic needs."(3)

The second issue is that the supply of goods demanded by the poor is not price elastic. These supplies depend upon a number of other factors, such as nature of land distribution, reforms and tenure, employment coefficients of techniques of production of these goods, and availability of inputs. Part of the reason these goods are not being produced in sufficient supply in the first place is the lack of responsiveness from social and economic institutions. Also, profitability in other sectors of the economy is liable to be far larger, particularly in view of the various explicit and implicit subsidies. All of this relates to the economic and political structure and institution. Their dynamics are like the turning of a wheel. As it turns, it throws a few people into the pile of the R society and the bulk into that of the P society. The effect of simple income transfers is to recycle some of the poor back through this wheel. As it turns, the majority still end up back in the same pile of the P. The World Bank strategy does not make any reference to changing this economic and political structure.

Accordingly, we find that the World Bank variant is basically a sophisticated formulation of CDS. It is not a variant of ADS.

ILO'S MINIMUM NEEDS STRATEGY

Recently, ILO articulated a "basic needs strategy" at the 1976 World Employment Conference in Geneva.(4) Basic needs are defined in terms of two elements. These include the certain minimum requirements of a family for private consumption (adequate food, shelter, clothing). These also include essential services provided by, and for, the community such as safe drinking water, sanitation, public transport, and health and education facilities.

The basic needs strategy must involve the active participation of the people in the process of decision making, particularly decisions that effect their living.

Obviously, there is no easy way to define the basic minimum of these goods and services. The minimum level will be determined by cultural, historical, social, economic, and biological factors. Thus the composition of goods and services will vary from country to country, from region to region in a country, and even from community to community in a region. The concept of the minimum is that of a floor and not a ceiling.

It is argued that these objectives of basic needs consumption be achieved through extensive employment generation. Actually, employment is considered a centerpiece, since employment generates both output and incomes and provides recognition to a person. It is also emphasized that a rapid rate of economic growth is an essential part of a basic needs strategy. The 1976 Conference made the year 2000 a target date by which most essential basic needs should be met in all societies.

The basic needs strategy idea certainly makes a frontal attack on abject poverty. This strategy, thus, easily forms part of an ADS. There are, however, two questions that are worth asking. Is the basic needs strategy complete by itself, or is it a part of a larger development strategy or plan? If it is only a part, how important is this part? This is an important question. For example, in the fourth and fifth five-year plans in India a minimum needs program was included. Unfortunately, it was so small and insignificant that from the point of view of the plan it was virtually irrelevant. The second question asks what the complementarities and conflicts in the insistence on "growth" and local participation are.

For the "basic needs" strategy to be successful, an answer to both of these questions is necessary. If a basic needs program is a part of a larger plan, it has to be a sufficiently large and important part of the plan in order to influence policy changes in the desired directions. Many times, policy changes only may not be sufficient. Also required is a change in economic and social institutions and change in the climate and ethics that affect the nature and composition of production. Yet, like the World Bank, ILO does not suggest, much less emphasize, these changes.(5) The necessary changes are in the institutions that generate employment and production. The most important relate to land, its distribution, tenurial practices, and feasibility of reforms. In many developing countries, even in the developed countries, basic needs programs will not work unless and until major changes are made in the institution of land ownership and tenure. The other important institution may be at the community level. Community level basic needs programs will be successful if, and only if, the problems relating to economic and political power are analyzed and resolved.

It is not clear why there is emphasis on "growth." In development literature, the term "growth" has become synonymous with CDS and

GNP. It is a buzz word which brings to mind all the philosophy, logic, and policies of the past 30 years. The whole thrust of "basic needs" is to change the composition of production, prices, and hence GNP. Much depends upon the concept of growth. If growth follows from job creation for the production of goods required to satisfy basic needs, growth and basic needs are complementary. On the other hand, if employment follows from growth which in its turn may arise from such activities as export promotion, urbanization, and pyramid building, where do the goods necessary for satisfaction of basic needs come from? Unless growth can lead directly to the availability — in sufficiently large numbers — of these goods through production or imports, there may be serious conflicts between "growth" and "basic needs." In these situations, a basic needs program will involve reduction in, and even elimination of, so-called growth.

PERSPECTIVES ON GANDHIAN DEVELOPMENT IN INDIA

In March 1977, there were elections in India. These were won by the currently ruling Janata Party. This party fought elections on the basis of a manifesto promising Gandhian Development Strategy which is an important variant of ADS.

The manifesto is more meaningful when taken as a whole.(6) Focusing on the economic part may be distorting, because the elements in the manifesto are not independent of each other. With this important warning, we give below the salient features of the economic elements.(7)

1) Institutional: end of destitution in 10 years; deletion of private property as a fundamental right; decentralization of the economy.

2) Production: production of basic goods for mass consumption; formulation of a full employment strategy; production by small-scale and cottage industries and reservation of spheres for these; development of appropriate technology for self-reliance.

3) Agriculture: higher priority for agricultural sector; institution of agrarian reforms; higher allocations for the rural sector; improvements in terms of trade between agricultural/rural goods on the one hand and industrial goods on the other.

4) Urbanization: development of a new village movement; reduction in urban-rural disparities; reversal of urbanization trends.

5) Inequalities: stress on Gandhian values of austerity; reduction in income disparities; recognition of wage earner's and women's rights.

In November 1977, the Janata Party formalized its economic policy statement. One would assume that this policy statement is derived from the election manifesto. This statement perceives the problems of India in terms of poverty, unemployment, and widening disparities of incomes and wealth. In this perception the policy statement is already somewhat removed from the manifesto. To solve these three problems, the policy statement advances a particular strategy of development. In their own words, "the new thrust of Janata Economic Policy would be growth <u>for</u> social justice rather than growth <u>with</u> social justice" (emphasis added).(8) This strategy of development involves a policy of maximization of output and employment, output and employment per unit of land in the agricultural sector, and output and employment per unit of capital in the industrial sector. Presumably, this strategy is contrasted with the maximization of output in the neoclassical development strategy. On the basis of this strategy, the program provides emphasis on the development of agriculture and small industry generally.

Some of the salient features of the program are as follows:

Agriculture:

A. (i) Forty percent of the public investable resources should be diverted to agriculture and rural areas.

(ii) Emphasis on these investments should be on quick maturing small-scale irrigation projects.

(iii) The availability – or nonavailability – of electric power has to be equalized between industry and agriculture.

(iv) The price parity between the prices paid and received by the farmers has to be established and maintained.

B. (i) Creation of jobs in the rural areas on such capital improvement projects as roads, tanks, wells, tree plantings, etc.

C. (i) There is a need for consolidation of, and a floor as well as limits on, land holdings.

(ii) The government has to develop and encourage a system of small independent farmers assisted by service cooperatives.

Industry

(i) The fields of production for small-scale industry need to be reserved so that large-scale industry will not add to its capacity in these areas.

(ii) The fields of production for national producers will be reserved by discouraging foreign producers.

(iii) Advantages between the small and large-scale production unit shall be equalized, in terms of raw materials, credit, technology, and marketing.

(iv) Large-scale production units will be encouraged to specialize in production for export.

(v) Large-scale production plants are not to be located in populated areas and must be kept away from cities with a population of one million or more.

Income and Wealth Gap

(i) Expansion in the large industrial houses should be restricted, even eliminated.

(ii) Progressive taxation of income and wealth should be imposed.

(iii) Reduction in the difference between minimum and maximum income after tax should be brought to 1:20 and eventually to 1:10.

Prices

(i) Reduction of inflation by control of money supply; incentive pricing for the production of basic goods; import and public distribution of basic goods; and effective control on trading.

There is a clear ADS component in the Indian economic policy. How effective it is remains to be seen. Much will depend upon the success of measures taken to change and develop economic and social institutions in order to make these responsive to the objectives of ADS.

CHINESE DEVELOPMENT STRATEGY UNDER MAO

In literature on development, it is customary to keep the socialist/communist countries apart. There is no doubt that there are fundamental differences between market and socialist economies. In the latter, the means of production have been nationalized and there are no parliamentary forms of government. However, it is our contention that these differences are irrelevant and immaterial to the strategies of development. In our view, socialist economies also follow what we have described as CDS and ADS. It is for this reason that we consider the experience of the People's Republic of China relevant.

When China gained independence, the strategy of development in socialist economies followed the practice of the Soviet Union. Soviet economic development placed heavy emphasis on investment, so that consumption was postponed year after year; and on centralized modes of production in terms of heavy industry, collective farms, big cities, and capital intensive technologies. However, Mao did not follow this strategy in China. There are different opinions among intellectuals of various persuasions regarding the nature, character, and success of development strategy in China. It is our contention that the development strategy in China, up to the death of Mao, has been a variant of ADS.(9) It contains all the elements we have listed in the last chapter, namely, redistribution and production of goods in favor of the poor in rural areas by decentralized means encouraging local initiative.

The elements of the Chinese development strategy may be listed as follows.(10)

1) Production is concentrated in a small number of material goods considered essential for all the population.

2) Major productive activities are very labor intensive.

3) A large part of production is agricultural and is carried on in communes where the production decisions are highly decentralized.

4) The development strategy has not encouraged industrialization as it has in other developing countries. If some industries have been needed, these have been set up and promoted.

5) Specialization is minimal.

6) The emphasis has been on the use of education, production, and technology to serve the people.

7) There has been little growth in, or of, cities.

As a result, this strategy has been able to eliminate both poverty and unemployment. It has also resulted in sharply reducing, if not completely eliminating, consumption differences and inequalities. This has been accomplished without any help from foreign aid since the Soviet withdrawal.

Professor John Gurly summarizes these achievements and places them in a larger development context.

The truth is that China over the past two decades has made very remarkable economic advances (though not steadily) on almost all fronts. The basic overriding fact about China is that for twenty years it has fed, clothed, and housed everyone, has kept them healthy, and has educated most. Millions have not starved; sidewalks and streets have not been covered with multitudes of sleeping, begging, hungry, and illiterate human beings; millions are not disease ridden. To find such deplorable conditions, one does not look to China these days but, rather, to India, Pakistan, and almost everywhere else in the underdeveloped world. These facts are so basic, so fundamentally important that they completely dominate China's economic picture, even if one grants all of the erratic and irrational policies alleged by its numerous critics.(11)

Part of the success of this strategy has been due to the fact that the economic and social institutions were fundamentally changed so that the new institutions, such as communes, were favorable to the objectives of the ADS. Soviet Russia had also changed the economic and social institutions drastically. However, the new institutions they created were not responsive to ADS goals; these were more suitable to other goals, basically those of CDS. If we compare the Soviet Union's and China's development strategies we find a number of differences and the following table outlines some of them. The table is self-explanatory.(12)

Table 9.1. Comparison Between Soviet and Chinese
Development Strategies

Categories/Emphases	Soviet	Mao
1. Modes of Production	Centralized	Decentralized
2. Production	Capital Goods	Goods for Basic Needs
3. Techniques	Capital Intensive	Labor Intensive
4. Education/Technology	Specialization	Nonspecialization
5. Rewards	Material	Nonmaterial
6. Location	Urban	Rural

Our conclusion from this comparison is that Soviet economic development is a variant of CDS while Maoist development is a variant of ADS.

OTHER VARIANTS OF ADS

The philosophy and strategy of development in Tanzania was laid out in 1956 in the Arusha declaration. It emphasized the development in rural areas and deemphasized the growth of cities and urban areas. The most important institution through which this process was to be encouraged is a "Ujamaa Village." The objectives of development and the nature of the Ujamaa Village have been analyzed and revised on the basis of experience. Most recently, President Nyerere defined the objective of development as the development of people. In his own words, "For the truth is that development means the development of people. Roads, buildings, the increase of crop output, and other things of this nature are not development; they are only tools of development" (emphasis in original).(13) However, people cannot be developed. They develop themselves. This is fully recognized. Accordingly, the concept of Ujamaa Village has also undergone a revision. An "Ujamaa Village is a voluntary association of people who decide of their own free will to live together and work together for the common good."(14) Hence the policy of Ujamaa Village is not intended to be merely a revival of the old settlement schemes under another name.

Obviously, an Ujamaa Village has all the elements of ADS. The production and distribution are based on local resources and local initiatives.   The process of decision-making is decentralized and participatory.   "Village" implies an emphasis on the rural and the needs of the poor.

It is difficult to gauge the success of the development experience in Tanzania. The indicators of success are different from the standard GNP, production, and price statistics.   The information on other variables is not easily available.

Guinea-Bissau gained its independence from Portuguese rule after a fifteen-year struggle, in 1974. Dennis Goulet has studied the development strategy of Guinea-Bissau. He defines it as one of the variants of ADS "in which distribution of benefits is more important then mere economic growth, and in which serious efforts are made to involve local communities in vital decisions affecting them."(15) There are three priorities in this development: agricultural development, improving human resources through education, and improvements in health and nutrition. The emphasis is on participation by the people. A new theory and practice of education is being developed which places more emphasis on "political" instead of "linguistic" literacy.

In the above sections we have listed some of the countries which have followed variants of ADS at a national level. However, virtually all countries aim to achieve some of the objectives of ADS, either nationally or regionally. For example, full employment is now a major goal of economic policy in virtually all developed countries. Japan, in the process of its own development, has relied heavily on small farms and tri-center areas. In Taiwan, production in agriculture has been encouraged by institutions which are in many respects similar to the communes of the People's Republic of China. One can find some examples in virtually every country.

10 Science and Technology in ADS: Appropriate Technology (AT)

CHARACTERISTICS OF SCIENCE AND TECHNOLOGY IN ADS

Science and technology are helpful in production. They help to increase the level of production. The technology most suited to the objectives of ADS must take into consideration the following facts.

A large part of the working population has no training or experience in operating machines of any kind. They are basically unskilled. All sorts of skills – engineering, management, marketing – are nonexistent. The availability of capital in the form of machines and other mechanical contraptions is scarce. On the other hand, unskilled labor is plentiful. The background required to apply scientific knowledge to production needs is limited. There is a shortage of training facilities and the basic source of learning is by doing.

These facts and the objectives of ADS determine the following characteristics of suitable technology.

The operation of such technologies tends to be <u>simple</u>. The maintenance and repair requirements also tend to be <u>simple</u>, and capable of standing rough use so that the people can learn by doing. They tend to liberate the human being from boring, degrading, excessively heavy, and dirty work, and they tend to be capable of producing basic goods needed by the poor people. The control of such technologies tends to lie in the hands of the people so that these technologies provide access to the poor people to solve their own basic needs. These thus tend to encourage individual self-reliance.

The technologies tend to be <u>labor intensive</u>. They tend to be available on a mass scale so that a large part of the population can employ these technologies. They also tend to use local resources in terms of material and energy which improve productive capacity on a sustained level of providing skills, encouraging capital formation, and R&D capabilities.

These technologies tend to be <u>small</u> so that they require minimal management and marketing skills, and the markets are usually not far

91

from the local areas. Accordingly, these tend to be located in local areas and they encourage decentralized modes of production and creative mass involvement.

The technologies are usually built on the <u>local technological traditions</u>, blending with, and enhancing, existing cultures. These also tend to use renewable resources thereby minimizing depletion of resources, and pollution. These may even tend to improve the environment. The technologies also do not hurt the poor and the weak by allowing them to be exploited; they tend to help the poor and the weak. By encouraging self-reliance and mass-participation these technologies tend to reduce inequalities of all sorts. In view of this, these technologies are sometimes called inequality-reducing technologies.(1)

We have used the verb "tend to" in order to suggest a process. This process is defined and developed by economic and social institutions as well as the "rules of the game." By "rules of the game" we mean reward and punishment systems. Without responsive institutions and favorable rules, these technologies end up being either nonexistent or inefficient.

The table below compares some of the characteristics of technologies relevant to CDS and ADS.

Table 10.1: Characteristics of CDS and ADS Technologies

Characteristics			CDS	ADS
A.	Output			
	(i)	income-level products	high	low
	(ii)	standardization	high	low
	(iii)	scale of output operation	high	low
	(iv)	satisfaction of basic needs	low	high
B.	Inputs			
	(i)	capital-worker ratios	high	low
	(ii)	skill requirements for operation and maintenance	high	low
	(iii)	marketing skills and levels	high	low
	(iv)	management skills and levels	high	low
	(v)	level of energy and resource use	high	low
	(vi)	labor productivity	low	high
	(vii)	capital (land) productivity	low	high
	(viii)	employment potential	low	high
	(ix)	use of local materials	low	high
	(x)	level of operation where people live	low	high
C.	Linkages			
	(i)	level of infrastructure	high	low
	(ii)	level of interdependence	high	low
	(iii)	ecological damage	high	low
	(iv)	inequality reduction	low	high

It is worth pointing out that the level of sophistication of scientific principles needed for the design of ADS technologies is not small. Some scientists contend that it is easier to design CDS than ADS technologies.(2) If these strategies are successful, the nature of technologies will have to be reexamined.

APPROPRIATE TECHNOLOGY

The successful implementation of ADS requires a choice by decision-makers in developing countries of technological paths congruent with the policy goals of such a strategy. The alternative path comprises not just suitable technology in the hardware or engineering sense, but supportive infrastructure designed to bend relevant choices by private entrepreneurs and government officials regarding specific importation, investment, employment, and distribution decisions in the direction of ADS goals.

The term "appropriate technology" (AT) is often used to describe the technological hardware and social software best matched to ADS and to local conditions in developing countries. At present there is no consensus about the term "appropriate technology." However, it is clearly recognized by its various characteristics.(3) In terms of material aspects of AT production, "appropriateness" connotes the use of renewable sources of energy and recyclable materials, minimum destructive impact on the environment, and maximum utilization of local resources. In terms of the modes of production, AT fabrication should take place close to the resource base, using processes which are labor intensive (capital saving), small scale, amenable to user participation and/or worker management, and located close to points of consumption. In terms of application, AT connotes assimilation with local environmental and cultural conditions. It does not overwhelm the community, but is comprehensible, accessible, and easy to maintain. Other phrases often used to describe AT are technology which is intermediate (lying in scale of sophistication between primitive and large-scale, contemporary technology), biotechnic (modeled on natural energy flows, maximizing thermodynamic efficiencies), soft (harmonious with the environment), and low cost (regarding price of inputs and products, and investment per workplace).

In its "hard" dimension, AT is a set of a large number of small, independent, diverse, and specific technologies. Some of these had been in operation and were superseded by large technologies as conditions changed. Others are still in operation or are the product of recent innovation or adaptation. Like all technologies, large and small, they have their problems. However, on the whole and at this point of time, AT provides a proper "fit" with factor endowments characteristic of most developing countries. It is also relevant to the ADS in its stress on maximizing the innovative capacity of individuals and groups within developing and developed nations, its sensitivity to the sociopolitical environment within which technological choices are made, and its direct application to dealing with basic human needs. On the latter issue, the link must be stressed between explicit national decisions to enhance the employment and income of the broad rural base of a developing country and to ensure more equitable distribution of income, on the one hand, and the choice of technologies responsive to employment needs and the production of goods both affordable and necessary in the light of human needs, on the other hand. In other words the ADS leads to choices regarding technology and processes of production which are labor intensive, relatively cheap, small scale in operation, decentralized in energy sources and location of production units, and so forth. Such feedback loops operate in this realm just as they do for the CDS/high technology/dual society path. In sum, AT qualities match the requisite characteristics of ADS technologies.

Several additional points must be made about the place of AT in the ADS in order to point out, and thereby remove, certain possible misconceptions. First, AT sometimes is viewed as obsolete, secondhand, inefficient technology intended by developed world elites to freeze the economic and technological status quo. In a similar vein, sometimes it is denounced as a tool of neocolonialism. This is a false perspective. If the notion of development as indigenous, innovative capacity is kept in mind, it is most imported technology that maintains the status quo of technological dependency. It is AT in fact which can help break this dependency.

Second, on the issue of efficiency, considerations about the viability of AT must be made in a context sensitive to social and cultural efficiencies, as well as traditional economic and technical criteria. AT is often most efficient in terms of long run social and resource costs. In view of its relevance to local labor and capital conditions, it may be even more suitable when considered in the light of such national goals as reducing unemployment, stimulating local entrepreneurship, maximizing use of available resources, and increasing national self-confidence and self-respect.(4)

Third, AT is sometimes viewed as a backward step to the Middle Ages. This is a misinterpretation of the desirability of adapting technologies used or developed in the past, which may not be presently used. Obviously, this view is false. It arises from a false conception of linear progress. The adapting of technologies, whether currently in use or used in the past, is in the highest scientific tradition. It is also the most efficient. In this sense, AT opens up the range and choice of technologies to match the objectives of ADS.

Fourth, in practice AT is not intended to be a complete replacement for the current range of technology imported by developing countries, nor could it be. The issue is to broaden the range of choices of technological means to fulfill the poverty-elimination goal, and to expand the technological spectrum available to decision-makers. There may well be cases in which sophisticated imported technology is the only tool to implement certain production modes. AT does not eliminate this end of the technological scale, but draws attention to the possibility of making better use of the labor and other resources at hand in developing countries. It is also conceivable that some AT programs, such as public health measures, must be implemented on a large scale.

Fifth, the fact that AT tends to be simple to operate is sometimes interpreted as if AT is scientifically naive. This is a misconception. In fact, the designing of AT is far more demanding of scientific theory and practice. It is so because AT involves the satisfaction of so many more constraints than the standard large-scale technology. This implies that the number of scientific principles that have to be mastered, and applied, are much larger. Some scientists argue that part of the reasons for the nonexistence of some obvious beneficial technologies for people may lie exactly in the complexity and advanced knowledge of scientific theory these ATs require.

Sixth, it is sometimes the view that AT is independent of social and economic institutions. This is not true. We have aruged that AT is defined by the objectives of ADS. ADS does involve, and imply, responsive social and economic institutions. Without such institutions the objectives of ADS cannot be achieved and AT loses its "A for appropriateness." We may repeat the slogan that "the best technology is a good society." Conversely, an inequitable society may co-opt the benefits of AT hardware for prevailing elites.

Seventh, and similarly, technological self-reliance does not imply a policy of total isolation from world economic markets, or the international political system generally.(5) It does, however, mean the possibility of a country delinking itself from particular markets or protecting its nascent industries in order to promote its internal development. Still, as noted, countries following the ADs will continue to engage in selective trade with developed nations, and more importantly, can engage in greater exchange of goods with each other.

Eighth, the use of AT does not imply a "no-growth" strategy. While this link may be suitably made for developed countries, in the Third World the issue is how to achieve growth with equity. AT is a means for doing so, for enhancing the kind of economic growth best able to provide for the needs of developing peoples.

Ninth, AT does not glorify drudge labor. It is intended to replace precisely the conditions of unproductive, burdensome work with more humane and efficient processes and tools. The concept of "right livelihood" is sometimes used to express the AT conception of work.

AT AND THE DUAL SOCIETY

In the literature on Appropriate Technology, appropriateness has not been sharply defined. As the practice of AT grows, so does the debate about what constitutes AT. Part of the reason for lack of a definition is that AT involves not only ideas but also practice. "For practitioners, working in the field, these terms, however defined, are perfectly clear and require little elaboration."(6) Another part of the problem is that AT is a movement which has been growing in recent years. As a result of this growth, a consensus on definition has not emerged. This is not peculiar to AT; it is true of all knowledge. In the early twentieth century there used to be various descriptions of what constitutes economics. There was a lot of debate about any particular definition. In the 1930s, this debate petered out and a consensus developed that accepted several definitions; the definitional debate was then left to epistomologists. Because of the smallness and growth of AT, its ideas are in ferment.

Now that AT is becoming a viable idea, it is attracting all sorts of people who offer different views of the term. Recently, the National Academy of Science entered into the fray.(7) This broadening of interest has intensified the debate about what constitutes AT. A NAS study has described AT so loosely and widely that many practitioners may not find it "perfectly clear." This has generated genuine confusion. There is, therefore, a need to define its concepts more sharply to discriminate between various technologies and their appropriateness.

Appropriateness is a derived characteristic. It is derived from certain objectives. Though AT is being practiced in developed countries and its practice is growing in the United States, the NAS study has developed AT concepts as if they are purely for the developing countries.(8) The basic issue arises from what objectives of the developing countries are considered. The study simply lists some of these objectives, without discriminating between them. Part of the reason for the lack of discrimination lies in the approach to developing countries – "us versus them." This is partly because the tools of analysis employed by the NAS study are not sharp. In Chapter 6, we suggested the concept of a dual society, made up of rich (R) and poor (P). This provides a useful and sharper tool. The NAS study has derived the appropriateness of technology from the objective defined and decided by the R society. Since the governments in developing countries are controlled by R society, the objectives that the NAS reflects are the objectives determined by the R society.

There are two related, but intellectually distinct, concepts of "appropriateness" that are relevant to the R society. One, which can be called TR1, is the technologies derived from the objectives of the R society, e.g., growth, power, prestige, and industrialization. The other involves technologies that are consistent with existence and growth of the economic and social institutions already established for the benefit of the R society, denoted by TR2. By and large, there are complimentarities between TR1 and TR2. However, it is possible that there may be conflicts between these two. The NAS study points out some of these conflicts. In an analytical framework, such technologies follow from CDS.

In international forums when governments formulate the need for technologies that they consider "appropriate," these largely tend to be those described above. This is particularly true of high technology developing countries, where the R society is well defined.(9) In the literature on international science and policy, there is a tendency to confuse the "interests" of the governments with the "appropriateness" of technology. So long as one defines appropriateness of technologies as above in TR1 and TR2, there is a confluence. However, the concept of "appropriateness" as discussed in AT literature is quite different, and may even be poles apart. This distinction needs to be made clear.

We argued in the last section on AT that "appropriateness" is derived from the objectives of ADS. The literature on AT reflects, by and large, such a relationship between technology and objectives. We also argued that ADS objectives are relevant not only to developing countries but also to developed countries. AT, thus determined, has meaning for all countries in the world. At a workshop, an attempt was made to develop a working definition of AT.(10) According to this definition, AT is concerned with processes and products. As a process it involves people in defining their own needs, designing and adapting products for efficient use within their own culture, environment, and economy. As a product, AT tends to be environmentally benign, small scale, decentralized and labor intensive; require minimal capital and skills; use local resources and talents; and enhance local self-reliance. The emphasis in this definition, which does not come out in many official reports, is on the people and their participation.(11) The term "people" does not means some abstract idea, such as a group (Asian, Americans, white, blacks) or a collection (consumers, workers). By people we mean individuals, human beings in flesh and blood. It is these individuals who separately and jointly define their needs. It is they who determine whether a particular technology is appropriate or inappropriate. It is they who not only accept and reject, but also control the technology.

It needs to be pointed out, and emphasized, that technologies, however appropriate, operate within certain social and economic institutions. They are not independent of each other. Reddy has called technology the "gene" of the society. If technology is the "gene," economic and social institutions are the womb. It is, therefore, necessary that the importance of responsive social and economic institutions is recognized. It is all the more important because such institutions either do not exist or have been in slow decay because of benign neglect. In the past 30 years, the institutions that have been developed in the developing countries have been those which are responsive to R society instead of the P society. Part of the reason AT is viewed as something which is beyond pure technology, and which contains elements of a movement, is the lack of such institutions and the need to create, redesign, and develop them. In the AT literature there is not as much emphasis on the nature, elements, and characteristics of such institutions as there has been on the nature of technology. It is our contention that such institutions are not consistent with doctrinaire socialism or doctrinaire capitalism or free market ideology. Instead they contain elements which may be interpreted from both ideologies.

CATEGORIES OF AT

We find two related, yet clearly different, subcategories of AT floating in the literature. For want of better words, we call these family-employing technology and community defined technology. These imply different economic, moral, philosophical, political, social, and welfare conditions. Thus, family-employing technologies are consistent with societies, even if they lack communities. Community defined technologies, on the other hand, presume the existence and preservation of communities.

Family-Employing Technologies

Family-employing technologies are those that encourage a family, and family friends, to employ themselves in the production of goods and services needed by the larger majority of the poor people. These satisfy at least three criteria. These improve the productivity of labor without replacing it; the control of operation of these technologies is in the hands of those who operate it; and these are labor intensive or capital-saving and encourage self-reliance and innovation. Obviously these have to be relatively cheap technologies, easy to operate and maintain, so that the majority of the people can obtain and use these. Also, these must use local resources in terms of inputs and energy.

Congressman Clarence D. Long has developed the idea of "light capital technology" which has commonalities with our idea of "family-employing technology." In his own words, the definition of "light capital technology" is as follows:

> Light capital technology should not be regarded as "primitive," "low," "unsophisticated," or "obsolete" technology. Rather, it is technology economical of capital. Producing a light capital technology that works, is culturally congenial, and is economic can require ingenious design and careful field testing.

> Light capital technology should not be regarded as synonymous with inefficiency or high cost. On the contrary, if done approrpiately, it should represent the least-cost solution by combining factors of production – according to their relative scarcities, economizing on capital wherever capital is scarce and expensive and labor abundant and cheap.

> Labor intensiveness is a necessary condition by which to define light capital technology, but it is not a sufficient condition, since even primitive or labor wasting technologies are labor intensive.

> Light capital is not defined by dividing the total cost of a project by some total of beneficiaries, especially where it is difficult to identify these beneficiaries and to measure their individual benefits. It is defined by a small amount of capital investment per worker using the capital, and preferably by small projects that can be managed by small entrepreneurs.(12)

Congressman Long has suggested $100 per worker as an example of the idea of "light capital technology." This amount and idea do reflect the concerns and spirit of AT.

The "family-employing technology" needs to be distinguished from contractual employment activity as in the old "put-out" system or as in the cottage or small-scale industry where the small-scale industry, to all intents and purposes, is integrated with the large industry – say in producing specific parts for airplanes – and operates like a wholly owned subsidiary of the latter.

Community Defined Technology

Community defined technology is one that is ideal for production at the community level. It presupposes the existence of a community and its preservation as a development objective. A community is defined by at least two conditions: People living together in a geographical area thereby sharing the pleasures and problems of the same natural phenomena; and a genuine human interaction between people in an area on a day-to-day basis so that people not only know each other but share each other's griefs, sorrows, joys, and rejoicings. We envisage genuine community as one of the social and economic institutions helpful for the success of AT in achieving ADS objectives.

A village in most of the developing countries can define a community or it may not. Ujamaa Village is an attempt to develop a community. On the other hand, suburbs on the outskirts of cities in the United States, and elsewhere, do not constitute a community. These do not satisfy the second part of the definition above. Vance Packard describes the idea of a community and the shortcomings of suburbs cogently.(13)

Production is not only a matter of capital, labor, and materials. It is equally determined by social relations and conditions. Such part of productivity is particularly affected by the existence, or lack, of certain community level infrastructure or facilities, e.g., drainage in the farming community. The provision of such infrastructure is sometimes impossible except by the community itself. For example, in an Indian village, a farmer can in no way construct drainage to drain water from his/her farm, if the drain goes through some other person's farm. Some of these community level infrastructures are very important, even fundamental, for production. Their nonexistence causes a lot of problems.

If these facilities or infrastructures do not exist, there are three solutions. One is a market or an individual solution. A particular individual invests heavy amounts and solves the problem for himself. Another is a community solution, and the third is a bureaucratic solution or a solution by the government from a far off place. It is generally recognized that the community solution is the least costly and most efficient solution. The cost curve of a community-defined technology is sketched in the graph.

A simple example will make this idea clear. Let us say there is a open sewage in a village. This open sewage provides a breeding ground for mosquitos and hence a seriously destructive effect on the health of the members of the community. The market or private solution is the employment of screening technology. Thus, everyone who can afford to installs screen doors, thereby reducing the impact of mosquitos in the

Graph 1

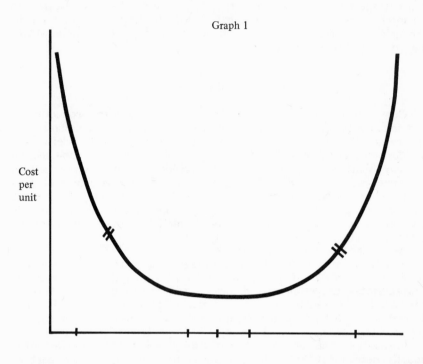

Cost
per
unit

Market Solution   Community Solution   Bureaucratic Solution
                                                     (including high technology)

house. The community solution is the use of underground drainage technology. The bureaucratic government solution may be spraying DDT, urban renewal, etc. depending upon the agency involved in solving the problem. It can be easily shown that community defined technology is least-cost and efficient since it matches technology to needs. Market and bureaucratic solutions, on the other hand, are extremely costly and grossly inefficient.

There is now increasing evidence that there are a number of "community defined technologies." Lovins makes his case for soft energy paths on the basis of the efficiency of the community defined technologies.(14) Brown and Howe have studied at length the question of provision of solar energy at the village level in Tanzania. Their cost calculations suggest that "at least five technologies now 'on the shelf' are, or soon will be, cost effective when compared with either diesel electrical generation or the existing electric grid in the case of the Tanzanian village."(15) Professor Minhas found that some of the issues raised in the Integrated Agricultural Development Projects basically dealt with community defined technology.(16)

The community defined technologies are defined, operated, and controlled by a community; they enhance the productivity of family-employing technologies.

REGIONAL AND NATIONAL INFRASTRUCTURE:
THE PLACE OF HIGH TECHNOLOGY IN ADS

In the discussion so far we have suggested that ATs are either small family-employing or middle-sized community defined technologies. On the other hand, in the past 50 years or so, such a variety of large or high technologies have been introduced, that it would be foolhardy to assume that some are not beneficial to the poor. There is no denying the fact that some of these large technologies are inequality-generating and harmful to the interests of the poor. The question, then, is about which should be selected and adapted in ADS. Looked at from another angle, AT involves providing technologies to the people. These technologies, in turn, have to be produced. Furthermore, the country has needs to develop regional and national infrastructure. Some of this infrastructure involves large or "high" technologies. Another question is about the place of high technology in ADS. It is continuously asked, particularly in the context of AT: When, if ever, is the high technology appropriate?

We can lay down a few criteria.

Since we have defined ADS as an improver of the welfare of the poor, the high technology under consideration must not lead to inequalities against the poor. The following example will make it clear. In every country, production of some foods is seasonal, e.g., apples in the fall in the United States; mangoes in summer in India; pistachios in Iran; cashews in Tanzania; and bananas in Central American countries. Let us look at the case of mangoes in India. In June, July, and August, there is a mango season in India; all the production takes place during these three months. Mango trees take years to bear fruit, so that the marginal cost of a mango in the summer is virtually zero, if not negative. Given no possibilities of refrigeration, all mangoes produced become the available supply, and the price of fruit falls. It falls virtually to zero, and the poor and unemployed people with no incomes have the time and labor to pick mangoes in season. They, thus, are able to get some mangoes which add to their nutrition. Now if we introduce the cold storage technology into this system, the owners of such technology are able to obtain mangoes during this summer period at near zero prices. Since refrigeration is a relatively large technology, it adds a new source of demand for mangoes in the area, and the price of mangoes goes up so that there are no more for picking. The supply that receives refrigeration is similar to that which was available to the poor people. Refrigeration keeps this supply and releases it for sale in the winter. The winter sale price has to be high because the demand price is high; and it must cover costs of transportation to and from cold storage, interest in the money tied up in cold storage, costs of maintaining the cold storage, and profits. The poor people, thus, cannot touch the mangoes in winter. All the cold storage technology has done is to take the food from the poor and give it to the rich, and the same argument holds for canning technology. Such technologies are inconsistent with ADS and are "inappropriate."(17)

We have defined ADS as improving the tendency toward self-reliance.   ADS discourages, and eventually eliminates, dependency relations.   The "high" technology must not generate dependency relations. Dependency follows from the satisfaction of basic needs. If poor people cannot satisfy their own basic needs, they become dependent upon those who help them do so. It is such incapacity that makes poor, unemployed, landless people dependent upon the landlord, because the landlord allows them an access to satisfy their basic needs. It is this dependency that provides the base for exploitation. Self-sufficiency, or lack of dependence, on the other hand, depends upon the frequency of the use of goods, a resource, or a service which in its turn is determined by the intensity of need.(18)

The intensity of basic needs is high, and the frequency of the use of goods that satisfy these needs is also high. Two examples will make this idea clear. The intensity of need for breathing is very high, and the frequency of use of air to breathe is also high. Similarly, the intensity of need, and the frequency of consumption, of food is high. The production of goods, that are frequently used and satisfy intense needs, has to be in the proximity of the people as consumers. In other words, such production has to be decentralized. Therefore, production of such goods should not be carried on in large and high technology formats which have to be located in regional and central areas. By the same argument, the production of goods (not frequently used to satisfy intense needs) by large and high technology are consistent with ADS.

No technologies — family-employing community defined, large and high — are consistent with ADS, and thereby appropriate if they do not cover long run social costs(19) and obtain high thermodynamic efficiencies by matching resources to needs.

CRITERIA FOR CHOICE OF APPROPRIATE TECHNOLOGY

An important question is: How is a particular technology to be chosen? In other words, What are the criteria for choice of technology? The choice of technology depends upon a large number of factors. It is determined by the decision-maker and the objectives the decision-maker decides to achieve. It depends upon the catalog of technologies available, availability of the information to the decision-maker, the availability of technology itself, and its capacity for successful adaptation to suit the particular needs and objectives. It is based on the availability of resources — in terms of finance, foreign exchange, etc. — and the existence of complementary infrastructure in the area where such technologies are to be employed.

One of the important questions about the choice of technology deals with the decision-maker and his/her objectives. In the real world of technology decisions, the decision-makers are government bureaucrats and executives of large corporations. Bureaucrats derive their objectives from political pressures and executives from policies for profit and growth. On the other hand, in the last section we identified two other decision-makers: family for the family-employing technology and community for the community defined technology. The question then is how bureaucrats and executives can reflect the interests of the family and the community. In other words, under what conditions would the decisions made by bureaucrats and executives be the same as those made by the family and community?

It is here that the role of responsive institutions becomes important. On the one hand, these social and economic institutions are able to transmit the hopes and needs of the people, or objectives of ADS, to the decision-makers. On the other hand, these institutions guide, influence, and direct the decision-makers to select technologies and policies in favor of satisfying the needs of the people and the objectives of ADS. Many problems in national and international meetings can arise from the lack of matching institutions.

We have argued that the purpose of these decisions should be to achieve the objectives of ADS. If decisions are made by the family for family-employing technology and the community for community defined technology, then the objectives of ADS will be achieved. We can then develop a criterion for the choice of technology from the objectives of the ADS. Formally stated, the technologies chosen should be those that lead to the achievement of the objectives of the ADS.

In the preceding sections we have discussed and defined the objectives of ADS. These may be listed as follows.

1) Provision of goods for basic needs of the poor.

2) Reductions in, and eventual elimination of, unemployment and underemployment.

3) Fostering of self-reliance and building on indigenous values.

4) Redistribution of income, investment, and wealth in favor of poor sections in order to bring about an egalitarian society.

5) Reduction in, and eventual elimination of, poverty.

Normally, it should be possible to choose technologies that achieve all of the above objectives. However, in case there is some conflict in the achievement of these objectives, a weighting scheme must be defined for the different objectives.(20) Weightings, by definition, are a valuation process and here reasonable scholars generally disagree. These should be obtained by discussions with the concerned people.

Many times, however, it may not be possible to either articulate the objectives clearly or to relate particular technologies to a particular set of objectives. In these situations, the criteria of technological choice based on objectives may become inoperative. There is, then, a need to develop alternative criteria for choice of a particular technology. In the previous sections we defined the characteristics of AT and the preferred characteristics of technology for ADS. Another criterion for choice of technology can be defined on the basis of the characteristics of technology. Accordingly, we suggest that the technologies chosen should be those that possess characteristics relevant to ADS. Congressman Long's amendment encouraging aid-giving agencies to select "light-capital technologies" for their activities in the developing countries is an example of this criterion.

In the section on AT, we argued that it is an important and useful medium for the achievement of the objectives of ADS. ATs possess some of the following characteristics in various combinations.

They tend to produce basic goods needed by poor people; to be labor intensive and amenable to family and/or community control; to employ local resources in terms of inputs of energy and materials; to encourage local initiatives and innovations; to be simple to operate, easy to repair and maintain; to involve minimal skills of management and marketing; and to cause minimal ecological damage.

Once again, normally, a particular technology should possess all these characteristics. They are all interrelated. However, if one has to choose among these, various characteristics will need differing emphases.(21) These should be determined by discussion with affected people.

Since these criteria are developed on the basis of matching technologies to objectives, it follows that these are also least-cost and efficient technologies.

AT: A SUMMING-UP

There is no single best definition of appropriate technology. Governments in developed countries tend to define AT – or rather define it away – through conceptualizations that are so broad that they could cover technology in general or that are too closely related to the objectives of the R sector. We have preferred criteria applicable to all countries and connoting some sense of the boundaries, however flexible, to AT. We have matched AT with three overlapping sets of characteristics: certain factor endowments (land, labor, resources, environment); certain modes of production and application, or processes and products; and a low range of cost per workplace. We have also related AT to several levels of scale: family-employing (household production); community defined (local infrastructure supportive of family technology); and regional/national high technology which does not leave the poor worse off or detract from their self-reliance.

The relationships between appropriate/inappropriate technology and society may be further clarified in Table 10.2. In Box A, there is a close link between CDS and conventional high technology. Similarly, Box D indicates a close link between AT and ADS. The situations that lead to conceptual confusion occur in Boxes B and C. Box B indicates that AT as hardware can certainly exist within the CDS society. However, the underlying institutions of such a society will not permit the pervasive use of AT to challenge the dual society status quo; enclaves of AT have little relation to larger social change efforts and are little more than a sop to those disaffected from the society. By contrast, Box C indicates that high technology can be used "appropriately" in ADS society because its institutions are oriented toward ADS objectives; in a sense, high technology will not be allowed to get out of hand. In the end it is the technology/society mix that determines the appropriate choice and use of technology.

Table 10.2. Technology-Strategy Mix

Development Strategy \ Technology	High	Appropriate
CDS	Close Links A	Possible, but Minor Role B
ADS	Possible, If Supportive of ADS Objectives C	Close Links D

11 Appropriate Technology in Practice

Given the concept and criteria of AT as we have expressed them, how has AT evolved in practice in both developed and developing countries? How have the connotations of "appropriateness" been translated in the field and in institutions seeking to support AT development? What problems or questions have arisen in the course of such development? We will explore these topics in this chapter by examining two dimensions of AT. These dimensions relate to our previous discussions of technology as both technique and package, and AT as process and product. AT may be considered a social movement and a collection of hardware and design alternatives presumably responsive to the ideology of that movement.

AT AS A SOCIAL MOVEMENT

We have stressed the fact that all technology carries a freight car full of cultural baggage. The trunks hold attitudes, values, and assumptions of society regarding the proper definition of the good life and the organization of productive means to achieve it. Technology symbolizes its own social impact. In this context, we have also stressed the particularly intimate linkages between conventional and alternative development paths, on the one hand, and certain types of technology and production processes, on the other. AT, then, is not only a set of tools with certain material characteristics, but it is a social movement.

As such, AT comprises a philosophy of technology and a loosely connected group of individuals and organizations that more or less share this philosophy, though with more widespread disagreement on the means for implementing it. The philosophy finds physical expression in the form of particular technologies labeled "appropriate" and political expression in the form of programs and tactics of relevant groups.

There is nothing particularly new about the ideas and practices behind AT. The growth of a popular movement supportive of the AT

106

perspective in the developed world is a phenomenon of the past decade. Ideologically, AT connects with a long tradition in the Western world critical of the social implications of industrial technology and the factory system in which it has been nurtured. In the twentieth century, this includes such individuals as Mumford, Ellul, Marcuse, Roszak, Gandhi, Illich, and Goodman. At its core is the belief that technological means have come to shape human ends, that technology has outpaced human efforts at controlling it.(1) The result is that man-the-toolmaker has become man-the-tool, engaged in alienating hierarchical work experiences, living only to consume, lacking all sense of community, and out of touch with environmental forces. This world view encompasses a deep skepticism of both the legitimacy and efficacy of technocratic authority, the benefits of urban life, and the commitment of capitalism and socialism to progress defined as bigger and better technology. It is apprehensive about the vulnerability of a society too dependent on centralized production modes and the impact of high technology on fragile ecosystems. A positive vision is also inherent in this perspective which correlates with our depiction of the goals of ADS. An alternative society is envisioned which would feature self-reliance, self-sufficient social units, human-enhancing labor, democratic decision-making processes, mutual aid, and harmony with nature. Thus, AT as hardware can be seen as the technical infrastructure of a decentralist society.

This critique of modern civilization and the technology on which it rests have always attracted adherents.(2) But the formation of a more widespread activist movement around these ideas had to await greater public awareness of what we have described as the various problems – inflation, diseconomies of scale, resource scarcity – associated with overdevelopment, as well as greater dissatisfaction with the products of affluence. Befitting its philosophy, and in addition to its recent origins, this movement is itself decentralized in organization and operation.(3) There is no party line or way to enforce ideological purity, no central national organization, no dominating leaders. In its consensual decision-making style and lack of generous funding, the AT movement shares similarities with, and draws many of its adherents from, the environmental, consumer, and feminist movements.

The activities and tactics of AT groups are most easily identified in the United States. However, they are springing up in all parts of the world, to such an extent that it is difficult to keep pace with these developments. The groups are extremely diverse and are usually based on their members' initiative, with no help from governments or large organizations (sometimes by choice, sometimes not for lack of trying).

In the United States, there are several particularly influential nodal points in the AT network. Think tanks and research institutes carry out policy research, innovation, field testing, and public education in AT hardware; some groups are university-connected, but most are not.(4) Journals and newsletters interconnect the dispersed AT community and report to it the findings of AT research, experiences in implementing AT programs, economic aspects of AT hardware, and critiques of governmental policies.(5) Leading luminaries travel the lecture circuit and add to a burgeoning AT literature, providing reassurance to those already converted that the cause is worthwhile and technically correct, and challenging the analyses of those who promote high technology.(6) Trade associations of AT manufacturing businesses engage in public outreach, promote research, and help lobby the government.(7) Short courses and workshops on learning AT-related technical skills are offered by private, and often nonprofit, educational firms.(8) Private agencies, some with religious affiliation, seek to extend AT information and technical assistance to developing countries, in cooperation with indigenous groups.(9) While not necessarily sharing a belief in all aspects of AT philosophy, government offices at all levels provide funding for AT research, development and public education.(10) All these elements come together at periodic conferences and fairs at which a wider audience can be proselytized, AT experiences shared on a personal basis, and plans worked out for lobbying, petition campaigns, and other ventures in support of AT.(11)

In developing countries, hundreds of equivalent groups exist, including private and academically affiliated research institutes, extension agencies, and governmental funding units.(12) Many of these groups tend to be initiated from the top down, rather than on the basis of grassroots initiatives, especially in those countries that have adopted ADS variants as official policy. Philosophically, the thrust behind AT in developing countries is more likely to be based on disenchantment with the CDS as we have discussed it in Part II, rather than a more sweeping indictment of the social effects of the industrial revolution. Sympathizers with ADS in the international development community are also less likely to be comfortable with the utopian and anarchistic elements that characterize much of the AT movement in developed countries.

At this stage in its evolution, two major issues confront AT as a movement. One involves the question of outreach to a wider public. On an impressionistic basis, the people who actively and/or ideologically associate themselves with AT are largely white, middle and upper class, and well educated. The leadership is also predominately male, perhaps reflecting the hardware aspect of AT. This feature of AT sometimes gives it the appearance of a current fad among those looking for the latest way to be ahead of the crowd – technology chic, as it were. This is ironic, considering the basic orientation of AT toward alleviating poverty, promoting employment, and generally improving the quality of life of the dispossessed in all countries. There is, then, a painful gap between AT aspirations and its need to draw into the movement minorities, blue collar workers and their unions, and all who are politically and economically disenfranchised from participation in the provision of their own needs.

Relationships with governmental agencies are another issue which embroils the movement in developed countries. From the debates carried out in AT newsletters and conferences, it seems that half the movement would like to avoid all entangling alliances with the bureaucrats, while the other half, more pragmatically, would like to divert in their direction at least some of the large budgets most governments allocate to subsidizing and promoting high technology. As we will explain below, the issue is moot at least from the government side, since various official programs are being established to further what agencies perceive to be AT. For AT people, then, the choice becomes whether to let such programs go their own way or whether to act like a traditional interest constituency working to ensure that its own representatives and views are included in the programs. The latter path entails the danger of cooptation, control, and dependency; the former, the danger of missing opportunities to utilize government resources in beneficial ways. There is no obviously correct answer here. In practice, AT groups are moving as their predilections take them.

AT AS HARDWARE

A major part of activities of AT groups has been concerned with the nuts and bolts of technology, which might be expected in the initial stages of the movement. The actual development of appropriate technologies has had a kind of catch-as-catch-can quality about it, highly dependent on the motivation of individual tinkerers, availability of markets for AT products, access to public or private venture capital, and access to the technical literature and experiences of AT co-workers. At least in developed countries, AT has evolved without systematic national planning or coordination; some would say this is the only way it could have emerged.

Numerous manuals and directories exist that give one a feel for the wide range of technologies available that are considered appropriate.(13) These technologies are derived from the modernizing of traditional technology, adaptation of high technology, and innovation in AT itself. Translating the criteria of appropriateness into material form is not a cut and dried process, because it is dependent upon the availability of local skills, resources, environmental conditions, and finances. Given major differences that can exist in these factors from place to place, a similar range of AT criteria or goals have been applied in both developed and developing countries, and in both rural and urban settings. As we have seen, these emphasize tools and production processes which are relevant to basic needs, relatively simple to operate, cheap, labor intensive, optimal with regard to local factor endowments, and amenable to local control, and which require relatively small capital cost per workplace. Presumptively, application of this framework leads to products whose quality is at least as good as products produced by high technology; whose cost reflects efficiency even by narrow economic criteria, whose environmental impact is more benign than that of high technology; whose thermodynamic efficiencies (where applicable) are significant; and whose long run and social costs are lower than those of high technology.(14) These points are best illustrated by a number of examples.

One calculation that has received widespread attention in AT literature is a comparison of two ways of producing 230,000 metric tons of nitrogen fertilizer a year. Illustrative of the potential efficiencies of dispersed, decentralized modes of production, the AT path would be to build 26,150 village level bio-gas ("gobar") fertilizer plants fueled by animal dung, at a foreign exchange cost of zero, at a capital cost of $125 million, and with 130,000 jobs created. A country could follow a high technology path, building one urban based, coal fired plant costing $70 million in foreign exchange and $140 million in capital costs, while employing only 1,000 people. The AT plants produce 6.35 million megawatt hours/year; the alternative plant consumes 100,000 megawatt hours/year of energy.(15) Similarly, an ILO report compares an oven imported by an African biscuit firm which cost $100,000 and created four jobs ($2500/workplace), with a tasteful and cheaper product which could have been produced from locally made brick ovens at a cost of $60/workplace.(16)

Theoretically higher production rates from complex technology, then, must be offset by other factors captured by the AT approach. These include improved income distribution because of higher employment, stimulation to local entrepreneurs, lower drain on foreign exchange, bringing jobs to the countryside, lower transportation costs of materials to production units and products to consumers, and simplicity of repair and maintenance operations. These variables come into play even, or especially, when the CDS path works on its own grounds. Thus, a developing country was able to produce 1.5 million pairs of plastic shoes and sandals a year using two plastic molding machines costing $100,000 and employing 40 workers. The sandals sold for $2 a pair. The cheap cost was an advantage, but the hidden costs included 5,000 local shoemakers out of work, drop in demand for local materials (leather, glue, cartons) not needed for the plastic shoes, and importation of materials that the machines needed.(17)

Nor are these considerations restricted to developing countries. The Metropolitan Sanitary District of Chicago has been faced with the problem of sewer lines which cannot contain the flow from frequent heavy rainstorms, causing back up of rainwater and sewage into rivers, streets, and homes. The District has planned and partially implemented a solution consisting of a large system of tunnels and reservoirs with an original price tag of $7.3 billion ($100 million/neighborhood in the region served or $1.6 million for each temporary job created). Moved by capital shortages and efforts of the local Center for Neighborhood Technology, the district has been considering small-scale alternatives, including retention of runoff through more parks and gardens, redesign of roofs, and reducing the amounts of sewage entering the system in the first place. Estimates in one area indicated that such strategies could achieve the benefits of the tunnel system at half the cost, not including opportunities for local job creation and the like.(18)

Employment opportunities arising from solar energy have come under intensive scrutiny in the United States. Particular attention has been paid to dispersed, on site uses of solar energy for heating and cooling, and conservation techniques (weatherization, insulation, storm window installation). A report of one such project pointed out that "on site solar technology appears to be more labor-intensive than contemporary techniques for supplying energy; thus, in the short term, the introduction of solar energy devices might create jobs in trade now suffering from serious unemployment. . . . the new jobs will be distributed widely across the country (and) should necessitate only simple retraining programs. . . ."(19) Solar equipment also need not be expensive if a medium of ingenuity is used. Thus, a group at Friends World College (New York) built a solar heater for less than $50 (from salvaged materials) which is suitable for sunny or partly sunny days.(20)

The fact that many AT projects require short or medium-term investment of relatively modest amounts, plus the community-enhancing properties of AT, makes community economic development efforts based on AT a prime antipoverty tool for blighted urban neighborhoods. At its best, community development results in the strengthening of local economic institutions, recycling of monetary resources within the community, and a wider share in ownership of and participation in modes of production relevant to the community.(21) Success in one area also tends to encourage spin-offs and other ventures by community groups, while further increasing investment opportunities within a community. Thus, community action agencies and development corporations in the United States and other countries have initiated enterprises based on such AT products as energy conservation, greenhouse construction, waste recycling, and housing rehabilitation.(22)

For example, there are 250 acres of vacant land in the South Bronx section of New York City which could be turned into parks and vegetable gardens. However, an acre's worth of topsoil needed for gardens would cost $13,000, which no community group could pay. Compost (humus) can be produced from organic garbage for half that cost or less. The Bronx Frontier Development Corporation has taken this approach, giving away a percentage of each year's product to community groups and selling the rest. Social benefits here extend to savings on transporting the waste used to landfill sites, which the city has been running out of anyway, as well as esthetic pleasure and learning experiences gained from urban gardening.(23)

Cooperative enterprises have also been linked with community development and AT in developing countries. The experience of the Arusha (Tanzania) Appropriate Technology Project illustrates this, as well as the amount of sophisticated thinking that often must be applied to coming up with simple designs. The Arusha windmill, developed by this group, is intended for water pumping, but can be used for grain grinding and other tasks. Engaging potential users in the design process, AATP came up with a windmill costing $1,925 (unassembled kit) that is durable, can be installed in two days, and transported to its site by a landrover. The kits are being produced by villagers of the Ujuzi Leocooperative, near Arusha.(24)

The field of AT energy supply for developing countries generally points up the importance of matching thermodynamic efficiencies with basic needs, an important lesson for the affluent world as well. Many of the power requirements of Third World villages can be satisfied without high quality electricity because all that is needed is low temperature on site applications, such as space and water heating, crop drying, cooking, and some industrial processes; even local electricity can be supplied by pedal powered generators, hydropower, and so forth.(25) It is this, plus the absence of nationwide electricity grids, that makes nuclear power so inappropriate for rural development, and even more so when one includes the foreign exchange costs and dependence on sources of enriched uranium that are entailed.

That AT is applicable at levels larger than the community has been shown by the introduction of ferro-cement for boat construction in Bangladesh. Ferro-cement is concrete reinforced with rod steel and wire mesh. In this case, it was intended to replace a simpler technology, wooden boats, because of the scarcity of lumber in a densely populated country heavily dependent on vessels for transportation. The ferro-cement has been manufactured from local materials, with the boats produced at a site that could be urban or rural, using supervised unskilled labor.(26)

In farming, even in the United States the presumed efficiencies of large scale, energy intensive agribusiness has been challenged and the benefits investigated of nonpetroleum dependent methods of agriculture using integrated pest management. The costs of synthetic fertilizers, pesticides, and herbicides are such that output and profits of organically run family farms are comparable with those of huge tracts owned by absentee corporations, with much less environmental harm and, as usual, with long-range social amenities.(27) One of the few empirical studies of the latter subject indicated widespread deleterious economic and social consequences for a rural town surrounded by corporate farms compared to one surrounded by family farms.(28) In similar fashion, several studies of farming in developing countries indicate that size of farm is not directly correlated with yield for modern crop varieties or propensity to adopt improved technology.(29)

At this stage in AT development, individuals and groups are searching for ways of integrating experience in various technologies to date. The logical step is the creation of an AT-oriented "new town." This is being done in the United States in the community of Cerro Gordo (Oregon) which, in the AT spirit, is being designed and constructed by its residents, and which features careful land use planning, alternative energy and waste treatment systems, banning of automobiles, and employment in local cottage industries.(30) On a larger scale, Tanzania is making similar plans for its new capital at Dodoma.

Issues regarding steps that can be taken to facilitate innovation in and use of AT will be discussed in Part IV. Here, several aspects of AT hardware, as it has developed, may be noted.

First, the technologies and design alterations that have emerged from attempts to apply criteria of "appropriateness" in the field are not fixed for all time. Constant experimentation is inherent in the AT approach as those who will use the technology continue to sharpen their sense of what is needed, as updated uses for traditional and even ancient technologies are discovered, and as AT research institutes continue to innovate new ways of using available materials. The point is that off-the-shelf AT is more pervasive and of wider variety than is commonly realized, rather than something that will only be available in the next century. Perhaps the bottom line is that AT as hardware is common sense engineering.

Second, some types of AT, particularly energy sources, can be scaled up in design to the extent that they are indistinguishable in deleterious consequences from the high technology they supposedly replace. This is a major reason that leads AT advocates to look askance at government funded programs for research and development in this area. In developed countries, speculation is rife about ocean thermal power systems, solar farms taking up huge chunks of land for electricity generation, and solar powered satellites beaming microwave radiation to Earth, while giant windmills and "power towers" (complex mirror systems beaming solar heat into generators) have been constructed. While all these plans may involve renewable sources of energy, none would be viewed as a victory for AT by the criteria we have discussed — community involvement is remote at best or may be shunted aside as in many nuclear power controversies; harmful environmental impact may be high; expense is high; and provision of electricity is not what is needed in the first place.

Third, and by way of contrast, it is possible for some tools of high technology to aid in research on AT. Thus, the New Alchemy Institute, a leading member of the AT network, has had funding from the National Science Foundation for the use of computers to monitor the intricate ecological interactions taking place in its habitat-greenhouse-aqua-culture-energy efficient structures ("arks"). Possibilities for lowering the costs of photovoltaic cells for on site electricity generation would strike some as a high/low technology hybrid. Space satellites may aid in communication of useful educational programs or resource observations necessary for land use planning, though here important questions would arise around degree of participation of users in preparing software or applying results.

Fourth, it cannot be repeated too often that AT as hardware is insufficient as an agent of social change in the direction that AT philosophy expresses. The social nexus in which the hardware is used is crucial. Thus, seen negatively, it is perfectly conceivable for AT to support suburban isolation in the United States and other developed countries. Two windmills in every garage reinforce the ethic of frontier individualism — self-reliance for myself, and the hell with my neighbors. That is not what is intended by the movement. Similar misapplication is possible in developing countries. Biogas plants, referred to earlier, tend to be owned privately in India, which means by relatively well-off farmers; cow dung, once free, has a market value. The landless and poor end up unable to use the dung for fuel and, to that extent, are worse off. In China, some 4 million plants are owned and operated by communities themselves, which benefit collectively from the methane gas produced.(31) In sum, AT is most true to itself when it is implemented not in technocratic isolation, but pursuant to a coherent strategy for social change.

12 Economic and Social Institutions in ADS

INTRODUCTION

We have suggested many times that the success of ADS depends, in an essential way, on the existence of economic, political, and social institutions that encourage, harmonize, and promote human behavior toward the achievement of ADS goals and objectives. Basic human needs have been felt since the existence of human beings. These have been satisfied since the early days of Adam and Eve. Over a period of time, institutions have been developed to satisfy these needs. These institutions are embedded in the cultures and traditions of people, areas, and regions. In the recent past, in view of the emphasis on CDS and its implicit policies to create uniform and standardized industrial societies all over the world, these cultures and traditions have either stagnated or followed a process toward slow decay. Instead, new institutions (consistent with CDS) have appeared, have been developed, and pushed.(1) We have in existence today all sorts of institutions. Those flowing from the various cultures and traditions; those created to advance objectives of CDS and industrial modes of production; and various combinations of these two.

The important question, in the context of ADS, is: Which of these existing institutions are consistent with, and supportive of, ADS objectives? This question breaks down into a number of subquestions: (a) Which of the CDS-defined institutions are complementary with ADS? (b) When do CDS-defined institutions become opposed to, and destructive of, ADS objectives? (c) Which of the institutions among the various cultures and traditions are purely contextual? (d) Which of the institutions in various cultures and traditions promote ADS? (e) Are the institutions, identified in response to the previous question, in a stage that allows them to be saved and adapted to meet current need? Or, are these beyond repair? What combinations of (a) and (d) are possible and when do these combinations become counterproductive?

The literature on these questions is scant. These questions have not even been raised, much less answered. To answer these, and ancillary questions, there is need for a major research effort.(2) In this chapter we do not provide an answer to these questions; frankly, we do not know the answers. We do not have in mind a complete design of economic, political, and social institutions responsive to the goals of ADS. We do not know how such institutions will look in their entirety. Our aim is a limited one – it is to outline some of the salient features of such institutions. We wish to point out basic principles by which some of the existing institutions can be analyzed. On the basis of such an evaluation, suggestions can be made for adaptation and improvements of existing institutions, and introduction of new ones.

FEATURES OF ADS INSTITUTIONS: THE CASE OF LAND

We have defined ADS objectives as "development of every human being" which involves, inter alia, the satisfaction of basic physiological needs. This requires that people can identify their basic needs, and have access to means by which these needs can be satisfied. It is a safe assumption that everyone can identify one's basic needs.(3) The basic issue, then, deals with "access to means." In a highly developed and monetized society, like the United States, money provides means to satisfy needs. If people can obtain money, they have the "access to means." Money in such societies is obtained by employment where one exchanges one's labor for money, and transfer incomes provided by the government in the form of unemployment insurance or benefits, food stamps, social security, and welfare payments. The society has provided two institutions for the "access to means"; a private wage-market, and an income-transferring bureaucracy.(4)

In developing countries generally, and poor countries particularly, both these institutions are nonfunctional. People are prepared, and willing, to offer their services for a wage. The industrial system in these countries is not able to provide employment to all of them. This is the meaning of the statistics that half the working force is under or unemployed. Furthermore, government resources are not large enough to transfer incomes to the two-thirds of the population who are poor. The issue of "access to means" therefore has to be analyzed more closely.

In poor countries a vast majority of people are poor and live in villages and rural areas. All the resources that these people have is their body labor. The only way these poor people can satisfy their basic needs is by using their labor to produce the goods and services they need. Unfortunately, labor by itself cannot produce food. To produce food, and other goods, one needs labor and other factors. The question of "access of means" then boils down to the "access" to these other factors. Most important of these factors is land. The majority of the poor people have no "access" to land for producing goods to satisfy their needs. This "lack of access" follows from the existing institution, in developing market economies, of landlordism. The institution of landlordism is defined when a person, a family, or a small group of persons, control (generally with, but sometimes without ownership) a vast area of land. Such a person is known as a landlord. By virtue of this control and ownership, the landlord determines which of the poor landless people will have "access" to land. The landlord also defines what use will be made of this land. In other words, if the poor landless worker desires to produce food for basic needs, this decision must meet the approval of the landlord. If the landlord, for whatever reasons, does not agree, the poor landless worker cannot produce food. The landlord, thus, controls the "access to means" for satisfying the basic needs. If the poor person has no alternative means to satisfy basic needs, he/she becomes dependent upon the decision, and willingness, of the landlord. This dependency relationship provides a fertile ground for exploitation, and the resulting humiliation, dehumanization, and all other indecencies.(5)

The objective of ADS is to "develop every human being." The impact of landlordism is to "degrade a human being." The institution of landlordism, thus, is diametrically opposed to ADS. In any society where landlordism is a living force, no amount of ADS is going to be effective. ADS in such a society is a farce. If a society desires to pursue ADS genuinely, it is of paramount importance that landlordism be eliminated completely.

In the past few decades many countries have made attempts to reduce the impact of landlordism. Ceilings on land holdings have been imposed; by law no one can own land beyond a certain size. Unfortunately, these measures have been purely legalistic. All these measures have done is change the identity of legal owners and landlords without in any way affecting the institution of landlordism.(6) Undoubtedly, the very large and conspicuous land holdings were removed. The interesting observation is that as the ceilings have been legally imposed, and legally obeyed, the number of landless and their proportion in the working force have also grown.

So long as there are sizeable numbers of landless people and they have no access to means to satisfy their basic needs, ADS requires that these people be provided access to land. The question is: Can such an access be provided? And if so, what institutional form will such an "access to land" take? We respond to the latter question first. There are at least two forms such an access can take.

Let all the land be redistributed, more or less evenly, among all the people willing and able to work on the land. This will provide the desired "access to land" to all the poor people. They will thus solve their problem of basic needs. The institutional framework needed for such a redistribution process could be similar to the income-transfer bureaucracy in the developed countries. One can, thus, argue for a land redistribution bureaucracy with a clear legal mandate to redistribute land. Redistribution must be emphasized. It has two dimensions: taking land from those who hold more than a certain amount of land; and giving land to those who hold less than the desired land. Thus, a bureaucracy with a legal mandate to redistribute land in favor of the poor would be consistent with the ADS.

The second form comes about if the land is taken from the landlord, and is given to the community for control and operation.(7) The "access to land" will, then, be defined if two conditions are satisfied. Every person willing and able must have a right to participate in community production, and the community production must be shared justly (equally, or in proportion to expanded effort, or in proportion to need, or a combination of these).(8) To satisfy these conditions, the political power in making decisions about production, distribution, and work allocation, has to be evenly distributed. This is possible if the community is made up of equals, preferably of people with moral commitments to justice and fairness. Furthermore, the community will need to have democratic institutions where decisions are made democratically.

Two conclusions follow from this discussion. Given the existence of landlordism, on the one hand, and landless people on the other (i.e., given the fact of unequal asset distribution), success of ADS and the solution of problems of poverty and unemployment require that economic and political power be evenly distributed by elimination of landlordism and redistribution of land in favor of the landless. If these conditions are satisfied, then two different institutions are consistent with ADS: a free-enterprise system with strict limits to land holdings regulated by a land redistribution bureaucracy, and a community-socialist production and distribution system with strong democratic institutions.

The question about the possibility, and practicability, of genuine land redistribution is a political one. It depends upon a number of factors, political strength of the people, the enlightened self-interest of the elites, and so forth. By arguing for it, we hope we are helping the political struggle in its favor.

HIERARCHY, COMMITTEES, INCENTIVES, AND TRUSTEESHIP

Production of material goods and administration of laws depends upon a combination of factors. The success of even small-scale operations is the result of joint efforts by a number of persons; such as members of the family. As the scale of operation increases, the number of persons involved also increases, though not necessarily at the same rate.(9) The end result is a joint product. Both these processes, production and administration, generate a number of decisions at various levels of the process.(10) In production, producers have to make decisions about what, where, and how to produce; how to combine various factors; what techniques to use; how to obtain resources (financial and nonfinancial); how to make persons work; what incentives to provide; how to sell, or dispose of, the output; and what part to distribute and what part to invest. The decisions related to administration are how to interpret laws and regulations; how and what information to seek and from where; how to obtain information; how to evaluate information; how to identify the offenders; and how to deal with offenders. These decisions translate into many other decisions, and sometimes they have to be made in a hurry and on the spot. If all persons involved in production and administration join in decision-making, there will be differences and even conflicts. There is a need for a mechanism by which such conflicts can be resolved.

The existing institutions have solved the problem of decision-making and resolution of conflicts, by development of a hierarchy and the formalization of hierarchical procedures. By and large, a hierarchy is defined by three layers, one on top of the other. All the decision-making power is concentrated in the top layer which is responsible for all the final decisions. In the top layer, the topmost person becomes the arbiter of the final decisions. The topmost person carries such titles as president, executive officer, chairperson of the board, and director. The topmost person represents the hierarchical unit, defines policies, and confirms all decisions. The bottom layer is, generally, denied any decision-making. They are the workers who just obey orders. The middle layer conveys the decisions and orders to the bottom layer and provides necessary information for decision-making to the top layer.

Hierarchies have flourished and reached their zenith in the form of bureaucracies of government and large multinational corporations. In the past 30 years, bureaucracies in virtually every government in the world and in every multinational corporation have grown at a fantastic rate.(11) Not only has the size of bureaucracy grown, but the number of layers between various offices has grown exponentially.(12) The interesting thing is that the bureaucracy of multinational corporations has also grown.(13) The hierarchical mode has been so prevalent that the international agencies like the UN, World Bank, and the Food and Agricultural Organization have maintained and developed large bureaucratic structures, even when they profess equality and peace.(14)

Hierarchies are based on the assumption of inequalities among people. People in the top layers are assumed to be smart, intelligent, and competent. By contrast, the people in the bottom layer are assumed to be ignorant, unintelligent, and incompetent. The reward structure is also based on these assumptions, and thus the people in the top layer are paid high rewards in terms of assets, incomes, perks, privileges, and work conditions.(15) Those in the bottom layer get the least rewards. The institution of hierarchy not only makes these assumptions, it also validates them by creating an environment in which such differences (imaginary or real) are maintained, accentuated, and promoted by introducing and increasing inequalities of economic and political power between the top and the bottom layers.(16)

Like landlordism, hierarchies in general have the effect of accentuating inequalities between people by concentrating economic and political power in a few and denying such power to a large majority. The greatest impact on the lower end is to degrade human beings, thereby impairing their development.(17) Remembering that the objective of ADS is to "develop every human being," hierarchies are not consistent with ADS; the larger the hierarchy, the more antithetical it is to ADS objectives.

Hierarchies do perform an important function — making decisions and resolving conflicts. The question is, if hierarchies are not acceptable, what other institutions can replace hierarchies? There does exist another institution. In the process of modern production and administration, there has been such an increase in the scale and complexity of decisions, that no human being is capable of making them all. This has gone beyond the human scale.(18) Accordingly, a new institution has been introduced, the committee. This institution has been grafted on the bureaucratic structure. However, in real decision making, it is already becoming the most important institution. It performs both the functions; it makes decisions and resolves conflicts.(19) Interestingly, the committee works best when the members are, more or less, equal.(20) If one or two members have much larger economic and political power, it breaks down and becomes a rubber-stamping process. Committees, thus, provide a substitute for hierarchies. Committees are based on cooperation while hierarchy is based on competition.(21) Committees can be appointed both by the people and the top layer of bureaucracy. This institution is also consistent with ADS objectives because committees elevate human beings. It satisfies a person's need for respect and self-esteem.

Hierarchy is explained as being based on the theory of incentives which follows. The more important the decision one makes, the higher the reward one deserves and higher the ladder of hierarchy one goes up. The topmost person in the hierarchy is expected to make the most important decisions.(22) It is assumed that important decisions require "work" – of some sort – and training. It is further assumed that any person can go up the hierarchy ladder by dint of "work" – the harder the work, the higher one goes. A hard working person does deserve plums. The theory of incentives may thus be defined as a positive relationship between decision-making and rewards; the more decisions one is responsible for, the more rewards. Hierarchy, and bureaucracy, is an institutionalization of this positive relationship.

We have argued that the objectives of ADS are opposed to hierarchy. Since hierarchy is an institutionalization of a positive relationship between decision-making and privilege, an institution consistent with ADS can be derived from the opposite of this relationship, – a negative relationship between privilege and decision-making. We enunciate this principle as follows. The more and important decisions one makes, the lower one's level of privilege and reward should be. For example, if one has the power to decide about the life and death of someone else, it is reasonable that such a person should live poorly. An institution based on such a principle will continuously equalize economic and political power.(23) As we understand it, this is one interpretation of the Gandhian idea of "trusteeship."(24) This is not a new idea. It has been practiced by Jesus, Mohammed, and other leaders. Its influence still continues, but the principle needs to be institutionalized.

MOVEMENTS TOWARD ADS INSTITUTIONS

In the previous two sections we have suggested some of the basic principles needed for the development of institutions responsive to ADS objectives. Development of institutions is a long process. It is like a large tree that needs time to take root; it needs a proper soil and favorable climate. The most favorable soil here is minimal inequalities in the concentration of economic and political power among people, regions, countries, and so forth. The most favorable climate is the existence and traditions of democracy.(25) Unfortunately, there are a number of countries, developing and developed, which have dictatorial regimes that deny basic human rights to people.(26) In such countries, ADS policies have little chance. The first order of business is to encourage movements away from dictatorship toward democracy at all levels.

Attempts are continuously being made to develop and strengthen responsive institutions. In Tanzania the institution of "Ujamaa Village" has been encouraged. This institution has been adapted from the culture and traditions of Tanzanian society. However, the climate in Tanzania is not yet fully favorable for this institution. The People's Republic of China has made attempts toward the development of a "commune" system. The commune is consistent with ADS objectives since it provides participation by the people at the local level. In the initial stages of the growth of Israel the kibbutz played a very important role. In India two institutions have lasted for centuries, the village community and the extended family system. The government has tried to introduce another institution from tradition, panchayati raj.(27) However, in view of the various policies in India favorable to CDS which have accentuated further inequalities in economic and political power, panchayati raj has not worked and the two other institutions have been badly damaged, hopefully not beyond repair. The AT movement in the United States has set up its own institutions, such as networks which hopefully and eventually will flourish into institutions of "conviviality" – to use a phrase from Illich.

There are a number of other existing institutions that have elements which could be quite consistent with ADS objectives. Credit agencies, banks, producer and consumer cooperatives, government and private R&D agencies, government-funding agencies, planning and policy agencies at various levels of government, information networks, trade associations, fairs, conferences, and affinity groups for direct action, contain large elements that are helpful to ADS. Given a favorable climate for ADS, these institutions can be adopted wholesale.

An important question is: What elements in the money exchange and profit systems are favorable to ADS objectives? Introduction of money relations makes exchange flexible, and it also leads to inhumanities. Up to a certain point, the profit motive produces needed goods. Beyond a threshold level, it becomes counterproductive. It is not clear what these threshold levels are. In recent years, interesting and extensive literature has come up regarding worker-managed and participatory production systems. It deals with humanizing the production system. Institutions that take these principles into consideration move in the direction of ADS.

13 ADS and Global Interdependence

INTRODUCTION

We have noted that developed countries use the rhetoric of interdependence to describe their perception of current international reality and to express a preference regarding how international relations should be shaped. This preference is not surprising. We have also discussed how conditions of "interdependence" actually cloak overlapping sets of dependencies between developed and developing countries and, in a mirror image reflection of the world stage, within developing countries themselves. With the exception of certain resource flows, these dependencies skew and reinforce the allocation of global goods — monetary, technological, cultural, etc. — in favor of the developed countries as a whole and, as beneficiaries of this arrangement, in favor of elites within developing countries. The losers in this game are the majority of the world's population who are poor and live in rural areas. With due allowance for marked differences in socioeconomic conditions, the present situation also discriminates markedly against the position of women in both developed and developing countries.

We do not believe that all this is the result of a conspiracy or the work of evil men, or that everything would work out if the whole world would adopt capitalism (United States variety) or socialism (USSR variety). Rather, the problems which we address stem from a misguided, though sincere, application of a development strategy at the wrong time and the wrong places, and with expectations that cannot be fulfilled. In turn, we have suggested an alternative strategy that emphasizes a different range of technologies and institutional mix. The logical question is the extent to which ADS may be expected to alter existing patterns of dependence and, in so doing, to give rise to relationships of genuine interdependence.

We recognize a certain unfairness in a procedure which compares the known shortfalls of the CDS as it has evolved in practice with the potential benefits that would ensue if ADS were widely adopted and

121

"correctly" implemented. Our remarks on the global impact of ADS are largely speculative, based on part on the vision inherent in ADS and AT philosophy and in part on practical experience to date. We are aware, then, that ADS itself, as it unfolds, may fall short of our expectations or may stimulate unanticipated consequences or may need re-evaluation over time in favor of still more appropriate strategies. Our remarks are made in the context of hope for the present and humbleness about human abilities to create the future as planned.

ADS AND DEPENDENCY PATTERNS

Masters and slaves are presumably interdependent, but that does not make interdependence a necessarily desirable state of affairs. Each "needs" the other, to be sure, but the content of the needs and ability to fulfill them are vastly different. Analogous relations exist among and within nations on the world stage today. To seek an international system which is less interdependent in the prevailing sense is not to work for a world of selfish, autarchic states engaged in the battle of all against all. Rather, in one sense, less interdependence of the kind we have means less dependency in certain key aspects of international relations – decoupling or delinking of developing countries from debilitating global networks dominated by the affluent.

The most obvious impact of ADS is on the economic and technological strands of the web of interdependence. The position of developing countries vis-a-vis developed nations and the multinationals in these matters has been recounted in previous chapters. A reorientation of these relations is inherent in the ADS from its stress on the self-reliant provision by a country of its own basic needs. A country following this path would seek to optimize development of its own productive forces to fulfill the fundamental physical requirements of its people, using local labor, skills, capital, and other resources. This does not imply autarchy if for no other reason than the fact that no country, even the most powerful, can totally care for itself, given geographic differences in arable land, climate, minerals, population, and the like. Some trade relationships will always be necessary. The point is that under strategies for self-reliance, a country would no longer base its development policy primarily on the existence of centralized, industrial modes of production. Therefore, it would not be under such pressure to import the requisite capital goods, industrial products, and trained personnel from developed countries. The importation of luxury goods that cater to an elite would also not be tolerated. A country would not have to earn the foreign exchange currencies necessary to pay for these imports by gearing its own output toward the needs and whims of markets in developed countries. It may still wish to do some of this, but the thrust of national economic efforts is in another direction. If ADS enables a country to escape the export oriented variant of CDS, the same can be said of the import substitution version. Here the nation would not be searching for ways to domestically produce the equivalent goods imported from abroad.

Breaking the chain of economic dependency does not stop with trade effects. If a country's ability to make effective use of domestic savings reduces its need of capital investment from abroad, to that extent it may be able to lessen its debts toward developed countries and multinational banks. Foreign debts in many developing countries have already approached the point where interest payments take up a major proportion of export earnings. Similarly, if providing nutritious food for its people is a country's primary agricultural goal and its land is not used to provide cash crops for export to developed countries, to that extent it may be able to avoid precipitous cycles in commodity prices that have bedeviled many developing countries. Foreign aid from developed countries which carries too many restrictions on the source of materials used, and which may depend on unpredictable swings in public opinion, may similarly be eschewed.

The links of technological dependency are also affected by ADS. There may be independent justification for establishing more scientific facilities and research and development capacity in developing countries to enable them to participate more effectively in global sociotechnical processes. We also recognize that exclusion of all but decentralized technologies is impossible – some basic needs may be appropriately provided by mass production methods, some centralized technologies may be peripheral to basic needs but useful for other purposes, and some countries may be so lacking in resource endowments that they require outside aid even for their basic needs. Granted all this, it is the purpose of ADS to highlight self-reliance in technology as the basis for lessening economic dependency. As we have seen, this entails taking advantage of the economics of community-level technology in order to maximize employment opportunities, entrepreneurship, and so forth.

Selective de-linking from high technology networks could, paradoxically, put developing countries in a better position to bargain over the terms by which useful technology not available locally should be imported.   That is, without the compulsion to acquire the latest technology multinationals have to offer, whether or not it will contribute to alleviating the needs of the poor, a country should be able to choose more carefully from among multinational packaging. Being able to set the financial and marketing conditions under which a country will purchase imported technology is as much a component of self-reliance as the ability to develop its own range of technologies unavailable, or available with burdensome qualifications, from the outside world.

ADS has implications for education and technical training as well. Once again the issue is not whether any students in a developing country should receive training in advanced science and engineering, or whether any students should be allowed to receive such training abroad. The issue is the basic orientation of a country's educational system. It has been characteristic of most developing countries to model their educational institutions after those of the former colonial country, to the extent of studying the history and culture of the latter in preference to their own. Conventional classroom education has enough inherent drawbacks of its own without imposing an alienating experience on top. Thus one can speak of appropriate education as much as appropriate technologies in energy, agriculture, and housing. Indeed, for the medium to match the message, education relevant to ADS would draw its subjects and content from the life experiences of the people to whom it is directed. This has been the approach in the more successful mass literacy campaigns carried out in several developing countries (Cuba, China). In our context, self-reliance policies would put a premium on familiarizing secondary school students with technologies, new or adopted, that speak to basic needs requirements, and on focusing the skills of advanced students on the often complex job of designing such technologies in cooperation with users.

Perhaps the deepest dependency pattern influenced by ADS is that of cultural assimilation, which supports the other components of dependence. This may well be the hardest relationship to break, for it goes to the roots of national identity and sense of self-worth. There are many vested interests built into assimilation habits. No culture can remain forever static or unaffected by communications and transportation networks which interweave the globe, but neither does the road to success require everyone to play ersatz Americans, or Frenchmen, or Portuguese. National authenticity can itself become a game. After all, wisdom is still a commodity to be found in all cultures. There are many ways to define progress and the good life, as well as the means of reaching them, as E.F. Schumacher was fond of pointing out in his homilies on Buddist economics or the Christian virtues. The value of these perspectives is not to hold developing countries in thrall to outmoded beliefs, but to provide alternative frameworks by which to judge the most effective paths to alleviating mass misery. ADS is intended to be receptive to and supportive of cultural autonomy in this sense.

ADS is not only relevant to asymmetrical bonds between nations, but also within them. If anything, this issue is even more controversial than the range of topics embroiling North-South relations, for it raises serious questions about the equity of internal political arrangements. However, the issue is unavoidable, particularly in the context of negotiations over a New International Economic Order. Whether sincerely or out of a cynical play to avoid further erosion of their own dominant status, developed nations have thrown back at developing countries the necessity for new domestic economic orders, so that a reshuffling of international resources will not reach the world's poor. For the elites of developing countries, then, ADS is a double edged sword — strategies of self-reliance can serve to buttress them against the more egregious forms of neocolonialism that go with CDS, but they also serve to diffuse power outward to social units not connected with the elites.

Why any dominant group would choose to adopt a strategy which in some ways undercuts its own status is a good question. The prospect may seem impossible if posed in either/or terms, but in reality things are not so stark. Many among the elites may perceive the connection between domestic moves toward ADS and international bargaining and, equally to the point, the internal dangers that arise when the P and R sectors of society are too widely separated, as Iran and other countries have experienced. Thus, some combination of enlightened self-interest and whatever political pressures can be exerted on them domestically may lead elites to view ADS as a palatable and necessary option. A thorough implementation of ADS, then, would affect patterns of dependence that exist within countries and that often reflect or have their roots in the center-periphery relations that hold internationally. For example, ADS implies that rural villages would no longer look to the cities for all sources of jobs and whatever other benefits trickle down from urban development. Rather, the villages would participate in the creation and sharing of the resources needed to fulfill basic needs. ADS taps local skills, know-how, and labor in order to situate jobs where the people are. The products of self-reliant work are then dispersed on the basis of both exchange and traditional nonexchange relations. In operation ADS pays deference to the informal entrepreneurship and technical knowledge that exist throughout rural societies. Expertise may still be tapped where relevant, but it would not be automatically assumed that consultants from a developed country or a traditional domestic educational institute have more practical advice to give than the people who live there. To this extent, ADS portends a lessened dependence of the village on the technocrat. Indeed, it is the latter who might seek out the former. Conscientiously applied, ADS would also include women in all phases of development planning and implementation, while stimulating the use of technologies that relieve the burden of hand cultivation, water bearing, and other sources of heavy labor which are traditionally performed by women.

THE WORST POSSIBLE CASES

The question is: Are there any "worst possible case" scenarios? One can imagine a situation where developed countries dominate innovation and markets in AT to the same extent that they do with high technology and its products. Here AT would become part of the same economic/technological dominence patterns that exist at present. The message might have changed, but not the process, with experts, academicians, and salespeople from developed countries converging on the Third World with the latest in participatory strategies, solar collectors, and appropriate flow charts. The AT movement in developed countries is well aware of this possibility, which it faces domestically in equivalent form. As we will note in the next chapter, there may well be areas in which developed/developing country cooperation can take place in an ADS context, but the possibility of developed country and multinational corporate cooptation of AT hardware and ADS jargon is always present.

Perhaps, more ominously, there are military implications of self-reliance. The ability to resist coercive threats from other countries and to carry on in the face of mass destructions in case of war was part of China's motivation for adopting economic and technological self-reliance under Mao. This is a defensive posture, but can one imagine a country feeling secure enough in taking care of its own needs to be tempted by aggressive provocations? It is also evident that self-reliance in military matters themselves (provisioning, transportation, some weapons) can enable a country to abuse it. However, people actively engaged in the daily tasks of production for basic needs and in planning for ADS goals are less likely to be wrapped up in the desire for military glory.

ADS AND INTERDEPENDENCE

The domestic and international changes that ADS portends disputably set the stage for genuine forms of interdependence. In this sense, we return to the connotation of interdependence as mututal reliance. Genuine interdependence, then, connotes more symmetrical forms of sharing world resources – material, financial, and intellectual – than exist at present.

For one thing, a lessening of trade between North and South could mean a rise in trade among developing countries striving to achieve self-reliance on a regional basis, as well as domestically. Thus, alterations in trade patterns within the developing world and between the latter and developed countries imply the emergence of new varieties of economic interdependence, rather than the complete end of such interdependence. There would still be room for national specialization on export goods more efficiently produced on the basis of a country's environmental and resource conditions – that is, for international divisions of labor. But in an ADS context, developing countries would not be using their agricultural land, for example, to satisfy developed country tastes before their own basic food needs were fulfilled, nor would developing countries be amenable to seeing their people used as a pool of cheap labor for the multinationals. As discussed in Chapter 10, there will still be a demand for high technology products which can be met from the industries of developed countries and some developing countries themselves. The reverse issue might also be posed. Wouldn't developed countries have anything to gain or learn from importing AT and personnel trained in AT from developing countries? ADS should enhance the possibilities for such learning experiences as developing countries gain confidence in policies for self-reliance and as developed countries come to acknowledge the relevance of ADS for their own citizens. Thus the notion of a technical assistance mission from Tanzania to solar energy groups in New York City, or a team of biogas experts from China aiding Midwestern farmers, makes excellent sense. In these ways, ADS may lead to more of a two-way flow in products, technology, and knowledge between North and South.

importance to developing countries of possessing the institutional means for making choices regarding ADS and AT issues. Following ADS, developing countries would work to create or improve mechanisms to carry out a number of relevant tasks. These include evaluation and selection of technologies suitable for acquisition, indigenous developing, and dissemination; location and encouragement of informal sources of AT (seeking out the involvement of AT users in research and field testing); research on removing market and financial barriers to ADS; and training and extension programs in AT. With this sort of institutional infrastructure in place (along with credit facilities, cooperatives, and the like), a country would be in a better position to participate meaningfully in regional and global negotiations related to ADS implementation. These entail efforts to de-package technological offers from ancillary services offered by the multinationals. Terms of trade for those products will still involve entering world markets and the establishment of cooperative programs within and outside of the developing world. ADS, then, strengthens interdependence by encouraging developing countries to improve their knowledge and policy base.

Finally, ADS implied a more heterogeneous world, in terms of the diversity of national cultures and the artistic means by which they find expression. ADS is premised on respect for and sensitivity to cultural diversity, analogous to its treatment of the nonhuman environment. Interdependence is not just a sharing of products and technologies, but of all the experiences and artifacts that human societies have developed in response to geography and history.

IV

Policy Issues
for an Equitable
World Order

14 Developed Countries and International Agencies

THE NEW AGENDA

We have argued for a shift in both the conceptualization of development and in the strategies for promoting development as they have been carried out over the last few decades. We have also argued for making the rich countries, as well as the poor ones, the targets of our preferred strategy, since conditions of overdevelopment in the former are intimately linked to conditions of poverty in the latter. Affluent nations need to alter their own resource use and consumption habits even as they address ways to respond to the varied and often conflicting demands from the Third World. Developed countries, in short, confront a new agenda, both domestically and internationally. The old agenda seemed clear enough. Formed from a mixed bag of motivations, including altruism, greed, and power politics, it stood for "more of the same." Developed nations, East and West, paid the piper, and the rest of the world could pick up the tune. What was promised was aid (in the form of loans, grants, technology transfers, and food) so that developing countries could follow the same path to material success, at a faster rate, as that followed by developed countries since the industrial revolution. Economically and politically, developed countries would remain in charge, dominating access to markets, resources, and technology desired by developing countries. Many in the developing world, especially in the elite sectors, bought into this agenda. It seemed a reasonable short cut to prosperity and, more subtly, continued the assimilationist process with which they were long familiar.

Now, however, a new agenda has emerged. The agenda is not altogether precise or coherent, but its essential message is the need for serious revision in prevailing development strategies and patterns of international relationships. This agenda is reflected in proposals for a New International Economic Order, directed toward the fulfillment of basic needs through Third World self-reliance. ADS means, in part, a lessening of the dependency of developing upon developed nations for technology, markets, and financing, or in other words, a lessening of the domination of world events by developed countries. The fact that much of the new agenda – the terms of the debate – has come from the Third World itself is one striking manifestation of altering patterns of interdependence. The natural question, then, is how developed countries will, and should, cope with efforts to implement ADS.

NONINTERFERENCE

The military capacity of developed countries gives them the ability to intervene, overtly or covertly, in the affairs of developing countries. There is a danger that the new agenda could give them the motive. To the uncertainties and risks inherent in any strategy for social change must be added the possibility that developing countries implementing ADS might awaken antagonisms in some part of the developed world. Hostile relations could result when developing countries, in the name of self-reliance, deny developed nations access to particular resources or commodities, or heavily control conditions under which multinationals can do business, if at all. When such policies are combined with Marxist rhetoric, developed countries may use the pretext to "destabilize" the offending government or otherwise take measures that make life uncomfortable for it (such as denial of credit, pressuring international bodies to halt aid programs, or financial and military aid to opposition groups). Intervention can also take the form of support for developing country governments that stick to CDS and face organized opposition committed to self-reliant alternatives. Since Vietnam, the United States has opted for verbal and economic support for governments in such cases, avoiding military action.

TOWARD THE CONSERVER SOCIETY

An important component of the stance developed countries take toward the Third World on ADS issues is their own domestic policy in this context. Indeed, the new agenda is being pushed within developed countries by concerned citizens who desire alternative paths for national development as much as it is being promoted in the international arena – and perhaps even more so. Some response is domestically required by developed countries who favor ADS in their bilateral and multilateral affairs and this follows from two aspects of our argument.

One reason for developed country domestic concern with ADS is that there are internal reverberations to many aspects of the international policies we have advocated. Changes in trade patterns, research priorities, laws regulating MNCs, technologies transferred, science and technology education, and other aspects of the new agenda all have domestic implications and applications. The reverse is also true. The extent to which domestic support exists for policies favorable to ADS internationally may affect the seriousness and effectiveness of developed country official backing for such strategies. Without domestic movements favorable to ADS both within their countries and as a direction for foreign assistance, developed country governments are not likely to enthusiastically pursue the matter vis-a-vis developing countries.

In working out their own alternative futures, developed countries are confronted by two opposing images connected by the phrase "postindustrial society." One social critic has given these images the labels of Hyper-Expansionist (with the telling abbreviation HE) and Sane, Humane Ecological (SHE).(l) The first version of developed countries futures is that expressed in the works of Daniel Bell, Herman Kahn, and others. Postindustrial here means "super-industrial," a society premised on the leading role of high science and technology, and in particular, on the information and service-based industries.

In such a society, citizens taken an even more passive stance than at present, accepting whatever material abundance comes their way, under the guidance of a benevolent, objective, and authoritative elite. There are no limits to growth, no problems that technology and expertise cannot solve. This image implies the continuation of developed country dominance internationally. It may be taken as the ultimate expression of the CDS philosophy, and as such appeals to those comfortable with the status quo.

The alternative postindustrial image fits our concept of ADS. Indeed, the logic of the ecosphere, the limitedness of world resources, and increasing pressures for international distribution imply it. The facts of overdevelopment demand it.

The rich nations' version of ADS may be called the "conserver" or "steady-state" society.(2) Such a society would put a clear premium on distributive justice, social equity, and environmental harmony. It defines the good life through value-goals which include local self-reliance, stewardship or husbandry of the land, frugal consumption, mutual aid, community-mindedness, craftsmanship, and an ecological ethic. Compared to the present, this society would presumably be less transient in its human relations, more cooperative, more participatory, less homogeneous because of the subsidiary importance of centralized institutions, and therefore more stable in the long run. The conserver society is analogous with the rich diversity of a complex, interdependent ecosystem. Physically, it would feature a constant stock of capital and population, with a relatively low rate of throughput to maintain the balance between input into the stock by birth and production and outflow from it by death and consumption. Optimum recycling of materials would be crucial to lowering pollution and resource waste.

The conserver society vision is best suited to deal with the facets of overdevelopment, guide the future of developed countries as a whole, both for their own sakes and for whatever good it might do in demonstrating their seriousness about ADS internationally.(3) We also feel that this model is capable of dealing effectively with the domestic consequences of developing country moves toward ADS/self-reliance paths. Realization of the conserver society must come from a combination of governmental and private action.

Second, there is the demonstration effect that can result when that part of the world which has served as a model for the benefits of advanced technology and maximum consumption begins to explore the consequences of and alternative to overdevelopment. Why should developing countries take seriously the developed country advocacy of ADS for them if there is no apparent connection between this strategy and the ways developed countries handle their own affairs? Obviously, developed and developing countries approach ADS from very different material positions, reflecting the global disparities we have described. But that makes developed country policies internally supportive of ADS all the more important. In the absence of such support, developed countries could hardly make the claim that the development path they have followed is morally and practically (in view of its impact on world resources and environment) a dead end — a part of the problem, rather than a solution to developing country ills. Without internal support, such a claim would justify the worst fears of developing countries that the basic needs approach is another paternalistic bid from the affluent. Just as we have argued that developed country and international aid agencies should express ADS goals in their own operation, so we would hold more generally that ADS begins at home.

In particular, it must occur in the United States. There is, of course, a nice irony here. It is the "American way of life" that most other countries in the world, even those in political or ideological disagreement with the United States, have sought to emulate. However, by the operation of some curious technological dialectic, it is in the United States that many have come to question the ultimate payments now falling due — economically, socially, and politically — for the efforts that have gone into making the United States the prime example of the CDS path. Perhaps the successful material results of the CDS have offered the opportunity and leisure for becoming aware of the pitfalls, but these are not less real and serious. Thus it may be that the most fruitful lessons the United States and other developed countries have to teach are the ways advanced industrial economies can develop themselves while maintaining, and even improving, their quality of life.

Governmental Policies

There are a number of steps developed country governments can take to support advances toward the conserver society. First, some governmental funding of AT is desirable to facilitate commercialization of such technologies at home and as a contribution to cooperative AT programs with developing countries. But there are clear problems. AT must compete with other R&D claims on the limited resources available from governments. Yet AT carries the burden of being an area of relative unfamiliarity to officials, with no traditional or powerful constituency, and with low public visibility compared to that of many high technology projects. Thus, complaints from AT practitioners about the relatively small support for decentralized solar energy sources have been endemic and characteristic for the field as a whole. These complaints are not simply over the level of funding in a total sense, but involve bureaucratic impediments and funding delays that may severely inhibit the interest of governmental programs of smaller AT entrepreneurs and community groups.(4)

Developed countries, then, need to provide expeditious funding and increased visiblity for AT projects, especially for small-scale R&D programs. We do not assume that all small research programs in general are AT oriented, or that large-scale projects are always inappropriate. The point is to get funds that are relatively smaller per award than government agencies are accustomed to awarding to qualified researchers, not necessarily connected with prestigious universities or think tanks, with minimum delay.(5)

Second, the "soft path" versus "hard path" debate over energy policies that has emerged in many developed countries is representative of a broader problem that extends beyond the types of hardware to be funded in one particular sector. "Hard" paths are generally congruent with the first connotation of postindustrial, involving capital intensive, centralized technological approaches to social problems. "Soft" paths are those compatible with conserver society goals. Hard paths are solidly entrenched in the fiscal and regulatory laws and policies, and in the institutional practices, of most developed countries. Purposefully or otherwise, these laws work to inhibit the use of AT and the adoption of conservation and environmentally-minded life-styles. Developed countries must carefully examine the economic and ecological costs of such biases, and the possible benefits to be gained by shifting public policy in soft path directions.

In the United States, one example of such policies and practices in energy includes large subsidies from the Federal government for centralized power sources.(6) Other policies include the reluctance of banks and other financial institutions to make loans for solar equipment; long-delayed tax credits for purchasing such equipment; energy prices that do not reflect full environmental costs or the cost of replacing nonrenewable fuels; obsolete building codes; and so forth.(7) In resource recycling and waste separation, barriers include artificially low waste disposal charges; less favorable taxation rates for industries using recycled materials than for those using virgin materials; and municipal laws prohibiting use of composttoilets.(8) Local laws may also inhibit experimentation with cheap building materials and passive energy designs. Similar policy biases can be found in health, transportation, agriculture, and other sectors. Changes in these legal and policy habits can be politically difficult, especially if short-run benefits are not apparent, but the liabilities of overdevelopment place their own pressure on governments to allow soft path technologies and design alternatives their rightful course.

Third, there is a need to reduce military spending. While localities in which military funds are spent may benefit, military spending tends to be inflationary for the nation as a whole because it does not contribute to consumer goods, among other factors. Several recent studies have also confirmed that spending on weapons creates fewer jobs than public outlays for social programs (housing, education, environment, etc.).(9)

Military budgets also absorb technical talents that might otherwise go into improving civilian productivity. There is an obvious connection between these two approaches to unemployment/inflation problems to the extent that money released from arms spending was channeled into soft development paths.(10)

Private Initiatives

No less important than governmental steps toward ADS, and to a certain degree the prerequisite for such steps, are ADS-related endeavors by private groups and individuals. Even while governments move slowly to adopt ADS policies, significant transnational links are being forged by ADS activists within and between developed countries. ADS groups play several roles. They demonstrate the effectiveness and benefits of AT hardware and software to fellow citizens; lobby governments for domestic and international support of ADS; provide funds to AT practitioners and research firms; and test out the viability of alternative life-styles consonant with reducing waste.

These activities are vitally necessary to persuade the public at large and governmental leaders of the reasonableness of ADS at home, and to test out nonhierarchical, humanistic organizational forms.

In addition to those directly involved in AT issues, the activities of three other social movements are particularly relevant to strengthening domestic popular support for ADS. First, peace conversion groups work to reduce military budgets and convert industries overly dependent on Pentagon largess to peacetime pursuits. The peace movement also includes antinuclear weapon activities, hooked up with those protesting civilian nuclear power and those who see AT as the alternative to centralized energy generation. A characteristic United States group here is the Philadelphia-based Mobilization for Survival, a coalition which takes under its wing a variety of antiwar and pro-social welfare local organizations. Several international conferences of activists from North America, Europe, and Japan have also been held by "mobe."

Second, in the past ten years, thousands of self-help groups have sprung up in the United States and in other developed countries. Motivations for such groups are rising prices for many basic needs products and a desire to participate in providing, individually and communally, for these needs. Thus neighborhood and community bodies are active in housing, energy, food, transportation, personal counseling, day care, and a wide variety of other endeavors.(11) Very little has been studied about the economic implicatons of these do-it-yourself or "do-it-with a litle help from your friends" activities. However, one such analysis on "household capitalism" – the home as a productive economic institution – cites a figure of $300 billion as the total of goods and services produced by households in 1965.(12) The significance of this movement is that it offers citizens of developed countries experience in a non-exchange-based economy, in which economic relations are not based on money, but on comity, trust, barter, and the like. Self-help groups, then, provide a strong thrust toward ADS.

The third relevant social movement for corporate accountability and better quality consumer goods is directly related to ADS. It works to change the shape and direction of corporate processes from within the developed countries where most MNCs are located. The Council on Economic Priorities and other groups in the United States have carried out competent studies on the impact of corporate spending, advertising, and production on the health and welfare of communities. These studies are often used by public interest organizations to bring pressure on corporations (through stockholder votes, government lobbying, direct action, etc.) to alter deleterious practices. These campaigns may concern corporate activities in developing countries, as in the case of withdrawal of investment portfolios from corporations with subsidiaries in South Africa, or the protests and law suits emerging in several countries around the provision of bottled milk to new mothers in developing nations.

MULTINATIONAL CORPORATIONS (MNCS)

The role of MNCs in technology transfer and penetration of Third World markets is controversial and not easily resolved. We see this issue in the following ways. As agents for the extension of developed country economies, and actions in their own right, MNCs draw developing nations into the full scope of dependency relations. MNCs are not going to disappear tomorrow − for better or worse, they are a fixture of the current international system. Not all products or activities of MNCs are necessarily inappropriate at all times and places.

The issue then becomes one of harnessing MNC talents in ways more congruent with ADS through positive and negative incentives so that developing countries may both decouple themselves from inappropriate dependency and engage in selective linkages with MNCs when suitable for promoting basic needs. To simply write off MNCs, then, strikes us as escapist; they exist and hard work is necessary to make their influence benign.

Several steps are possible. First, the ability and willingness of MNCs to produce and/or market appropriate technologies needs exploration. Important issues here involve possible existing examples of AT products from MNCs (e.g., farm implements from John Deere and the Singer Corporation); formation of MNCs among developing countries around AT products; and possible reactions of MNCs to alterations by developing countries of economic policies in favor of intermediate and low-scale technologies.

Second, developed countries must examine how their own policies work to subsidize the export of unnecessary and harmful technologies by MNCs to developing nations. On the negative side, developed countries need to look at the possibility of overt steps to prevent the export of hazardous technologies, including toxic substances (e.g., pesticides) and nuclear plants. Precedents for export control lie in governmental regulations of military equipment sales and in efforts led by the United States to restrict foreign sales of nuclear-processing technology. Such moves would be clearly discriminatory against developing countries if unmatched within developed countries themselves. There is a good case for cutting back or abolishing the stimulations to foreign investments offered by programs like those of the United States Overseas Private Investment Corporation, on either ADS grounds or because MNCs hardly need governmental support in the first place. The possibility of extending domestic laws on restrictive business practices, environment, and consumer protection to corporate technology transfers might also be considered.(13)

On the incentive side, developed countries might encourage MNCs to tailor their products and manufacturing processes to developing country needs through adaptive engineering. For example, one scholar holds that AID could request firms investing in developing countries to develop labor intensive plant designs, for whose expenses AID would pay if the design were not used. AID could also make grants to equipment manufacturers for development of AT and to consulting firms who broker investment openings in developing countries with United States businesses willing to use AT in response to such opportunities.(14)

Third, at the international level, discussions in the UN and other international organizations have centered around codes or guidelines to regulate the conduct of MNCs in their dealings with developing countries. Two drafts regarding general codes of conduct for MNCs have been written in response to Third World demands for greater control over MNC activities. The draft from the organization for Economic Cooperation and Development (OECD), as might be expected, stresses the investment climate, while that from the UN stresses development issues.(15) Sponsored by the UN Conference on Trade and Development (UNCTAD), another effort is underway to write an international code of conduct on the transfer of technology, which would attempt to lower the cost of such transfers to developing countries, remove barriers placed on recipient countries by MNC contracts, encourage indigenous research and development, and so forth.(16) To the extent that such codes are intended to facilitate the transfer of high technology under better or fairer terms, they fit comfortably within the CDS. The proposed codes may be useful as a means of bringing pressure on MNCs. Otherwise they appear to do little for ADS and the goal of self-reliance. That is, they may have the effect of ameliorating the more reactionary technology practices of MNCs without affecting basic dependency relations.

FOREIGN ASSISTANCE

Most developed countries operate foreign assistance programs on a bilateral basis, and make contributions to multilateral programs run by the UN and its specialized agencies. Some developed countries, such as Canada, Netherlands, and Sweden, have placed heavy emphasis on facilitating ADS within their aid programs; in others, the experience is mixed. The United States falls in the latter category.

Throughout the early 1960s, United States foreign assistance was premised on the conventional development strategy, with the underlying motivation of securing Third World allies against internal and external subversion. As the CDS came under question, development was recognized as a legitimate goal in itself. In 1973, Congress gave explicit mandate to a redirection of United States foreign assistance efforts by requiring that assistance focus "upon critical problems in those functional sectors which affect the lives of the majority of the people in the developing countries," with "highest priority . . . to undertakings submitted by host governments which directly improve the lives of the poorest of their people and their capacity to participate in the development of their countries."(17)

Development assistance programs such as those run by AID are not the only way the United States influences financial and technological flows from itself to the Third World. The United States supports foreign investment by American companies through varying subsidies and incentives. These include low-interest loans to developing countries for the purchase of United States technology made by the Export-Import Bank, tax deferral and credit laws applicable to overseas investments, and investment insurance against war risk and expropriation and loan guarantees offered by the Overseas Private Investment Corporation (OPIC), a governmental body.(18) Intentionally or not, such programs encourage the transfer of capital intensive, high technology to developing countries, with presumed benefits for Americans employed in the aided industries.

Among the international specialized agencies, the World Bank plays a leading role. In fiscal year 1978, development loans from the Bank, its affiliates, and the regional Inter-American, African, and Asian development banks totalled $11 billion, and disbursements, $5 billion. The United States share of contributions averaged 25 percent. The World Bank traditionally emphasized capital intensive, large-scale, urban-oriented technology in its projects and a trickledown theory of development.

Besides official development aid from governments and intergovernmental organizations, hundreds of private, nongovernmental associations, national and international, are engaged in assistance endeavors. These groups include OXFAM, CARE, Church World Service, Volunteers in Technical Assistance, Ford Foundation, and the like. Generally, their budgets are small compared to official agencies. Making a virtue of necessity, they have taken pains to work closely with indigenous populations in projects, and target their programs more selectively than is usually the case with large governmental efforts.

Whatever the rhetoric, three sets of problems can be identified with the kinds of programs we have outlined.

First, there is a kind of policy schizophrenia among programs and organizations involved in the aid game. It involves attempts to facilitate self-reliant development of developing nations. This is evident in both the United States and the UN. First, one school of thought sees any form of foreign aid, whatever its basic thrust, as a more or less disguised brand of neoimperialism, puposefully designed to draw developing countries politically and economically close to the West.(19)

Whether intentionally or not, this is the result when aid programs strengthen vested interests in developing countries who base their power in part on the domestic manipulation of aid and investments, or tie grants and loans to purchases in the donor country, or promote technologies needing spare parts and expert maintenance best obtained in developed countries.

Second, whatever their efforts to help the poor improve their own lot, most developed countries have programs for military assistance and commercial weapons sales that work to stabilize the regimes of Third World allies who may be unresponsive to strategies for self-reliance. Third, while some agencies are furthering alternative strategies as they perceive them, others are supporting CDS and high technology paths for developing countries. Thus a developing country facing serious debt problems may find itself required by its creditors (multinational banks and official loan agencies) to come under the oversight of the International Monetary Fund (IMF) as a condition of further aid or debt rescheduling. The IMF is another specialized UN agency that makes loans of foreign currencies to countries in need, contingent upon a country's adherence to strict deflationary policies and other efforts to further conventional means of growth. Whether ADS meets the IMF world view of how to strengthen a country's economic position is problematical. Analogously, in the United States there is an inherent tension between AID programs to promote ADS and programs of other bodies to promote the role of large corporations in investment and business operations abroad. To increase the ability of multinationals to penetrate Third World markets is to increase dependency relations between developing and developed countries, while the intention of ADS is to break such relations where they work against the best interests of developing nations.

A second set of problems arises at the bureaucratic/institutional level. There is a question of the commitment of at least top-level officials in the United States and UN aid agencies to ADS and of the relevance of programs implemented under their legislative or self-defined mandates. The basic thrust of AID, the World Bank, and UNDP still emphasizes high technology and CDS variations. A recent study on the impact of development spending on environment states that from 1972 to 1977 loans for energy by the Bank, its regional colleagues, and UNDP focused almost entirely on large-scale projects intended to deliver high-grade electricity for urban and industrial uses, and on electricity grids. Programs are similarly skewed in the sectors of human settlement, water resources, and forestry.

A more fundamental issue is that posed in Chapter 12 regarding the hierarchical, authoritarian nature of bureaucratic institutions, in which privileges are awarded on the basis of one's decision-making authority. We have suggested the trusteeship and committee concept as more compatible with ADS ideals. However, the UN, World Bank, and governmental agencies (and corporations) have a difficult time breaking from the traditional mode. Thus, there is always a gap between decision-making processes characteristic of such bodies and the ADS they may be trying to promote somewhere else. Without experiencing the egalitarian, participatory style necessary to ADS in practice, officials of the agencies will be faced with a conceptual and psychological barrier between their own work and the people for whom it is intended. This barrier may be more or less serious for the effectiveness of ADS-oriented programs, particularly when combined with an agency staff whose training is based largely on Western models of development, and toward expensive technology.

SOME POLICY ISSUES

A range of policy options and directions have been put forward in this chapter. Some are listed below.

1) At the domestic level, developed countries should pursue policies that aid their transition to conserver societies and economies. Such policies include expeditious and increased funding for AT projects; examination of and change in laws and regulations that bias technological and eocnomic choices toward "hard" paths; and the pursuit of full employment. Nongovernmental initiatives are also very important in convincing the mass public of the viability and benefits of adopting alternative life-styles which lessen the burden developed countries put on global resources and ecosystems.

2) At the international level, developed countries are to refrain from interfering through military or other coercive means in developing countries attempting to implement ADS. In co-operation with international aid agencies and developing countries, developed countries should adopt foreign assistance programs that maximize participation by recipient countries and their citizens in projects; that strengthen developing country capacities to analyze and choose from competing technologies those best suited to their needs; that enhance the exchange of information and experience about AT; that further research development in AT; and that promote education and training in AT and ADS concepts and practices. Developed countries also should ensure that their overall fiscal policies do not subsidize the export of inappropriate technologies to developing countries and should stimulate multinational corporations to adapt their projects to develop-ing country needs.

15 North-South Dialogue: CDS or ADS?

NEW INTERNATIONAL ECONOMIC ORDER

The United Nations came into being after the Second World War. In its first meeting it had 51 member states. Thirty of these member states were from developing countries. Twenty out of the 30 were from Latin America.(1) As erstwhile colonies became independent states, and the colonial countries wanted greater influence in the UN, membership in the UN grew. By 1977 more than two-thirds of the member states in the UN came from developing countries. Since the UN is based on the principle of one member, one vote, this has changed the character of the UN. It has rationally changed the tenor and tone of the debates. The most important elements in this debate arise from two issues, the desire of developing countries to develop, and the wide gulf between the developed and developing countries.

Around the same time as the beginning of the UN, the Bretton Woods Conference established two institutions: the International Bank for Reconstruction and Development (World Bank), and the International Monetary Fund. These two institutions formed the centerpiece of the international economic system. Instead of the system of one member one vote, voting in these institutions was defined by financial contributions. The developing countries have been admitted to these institutions; however, it is the developed market economies who have maintained complete control in the functioning of these institutions. It is only in the past few years that some Arabian petroleum exporting countries, with their unusually large holding of international currency, have had influence in the policymaking of these institutions.(2)

Development contains a large number of economic issues. There has been a general feeling among the governments of the developing countries that the international economic system is not helpful in their development efforts. They have maintained that their export earnings have suffered uncertainties and large fluctuations. Since the exports were mainly crops or raw materials, these have been seriously affected by synthetic substitutes. The prices of these goods, in the international market, have been low and subject to fluctuations. In other words, they have not been receiving a fair price for their exports. On the other hand, the prices they had to pay for the imports of manufactured and capital goods from developed countries have been unduly high.(3) Since imports have been considered necessary for development, in the manner of CDS, they have experienced foreign exchange shortages which have been thwarting their development. The governments of developing countries have, therefore, focused debate on these issues and sought financial aid from developed countries, on the basis of past exploitation and future developmental needs, trade stabilization, commodity arrangements, foreign exchange liquidity, and so forth. These debates led to the declaration of a "development decade" and to the establishment of the United Nations Conference on Trade and Development (UNCTAD) in 1964 with the purpose of improving the trade of and securing aid for, developing countries.(4) UNCTAD, like the UN, is a debating body. At best it can help in the formulation of opinions. Real resources lie with the World Bank, IMF, and in bilateral trade and aid relations. After UNCTAD, the UN upgraded another organization, United Nations Industrial Development Organization (UNIDO). The UNCTAD Conference every four years has accentuated this debate.

Since UNCTAD's establishment in 1964, two new elements have been introduced in this debate. One, a number of developing countries (which we have classified as high technology developing countries), has established a large industrial sector.(5) This industrial sector has been producing manufactured and capital goods. But since the internal market for these goods is limited, the maintenance and expansion of this sector depends upon export.(6) These countries have realized that the market for such manufactured goods exists mostly in the developed countries. However their access to these markets is limited, for various reasons such as tariffs and trade preferences. The needs of these countries, thus, have introduced a new issue in debate – the access to markets in the developed countries.

Two, around 1970 the petroleum exporting countries realized that the continually increasing demand for petroleum in the developed countries will exhaust their resources in the not unforeseeable future. In some of these countries, petroleum is the only resource. Thus, in 1973, they joined together and formed the Organization of Petroleum Exporting Countries – OPEC. They have had two results. Overnight they raised the price of petroleum many times.(7) They have been able to maintain this price for five years now, and this has resulted in a major transfer of resources from the developed countries to OPEC. It is already affecting the international economic structure.(8) Another result was that Arab members of the OPEC declared an embargo on the export of petroleum to countries supportive of Israel in the Israel-Egypt war. The embargo was successful and revealed how vulnerable many of the developed countries are.

It is against this background that the United Nations General Assembly, at a special session in April 1974, declared the establishment of a new International Economic Order. Partly, this declaration reads,

We, the Members of the United Nations, Having convened a special session of the General Assembly to study for the first time the problems of raw material and development, devoted to the consideration of the most important economic problems facing the world community, Bearing in mind the spirit, purpose and principles of the Charter of the United Nations to promote the economic advancement and social progress of all people Solemnly proclaim our United determination to work urgently for THE ESTABLISHMENT OF A NEW INTERNATIONAL ECONO-MIC ORDER based on equity, sovereign equality, independence, common interest and cooperation among all states, irrespective of their economic and social systems which shall correct inequalities and redress existing injustices, make it possible to eliminate the widening gap between the developed and the developing countries and ensure steadily accelerating economic and social development in peace and justice for present and future generations.(9)

This declaration has seven clauses. The most important is the fourth which lists 20 principles for the founding of the New International Economic Order.(10) Basically, the principles deal with just and equitable prices, assistance, and transfer of resources and technology to developing countries, and so forth.(11) The declaration ends with the seventh clause that reads "This declaration on the Establishment of a New International Economic Order shall be one of the most important bases of economic relations between all people and all nations."(12) The General Assembly, at the same session, also adopted a program of action.(13) It deals with such issues as problems of raw materials and primary commodities related to trade and development, food, general trade, transportation and insurance, the international monetary system and financing of development industrialization, transfer of technology, regulation and control over the activities of transnational corporations, promotion of cooperation between developing countries, assistance in the exercise of permanent sovereignty of states over national resources, and strengthening the role of the UN system in the field of international economic cooperation.

NIEO is based on the fact, and assumption, that there is a serious inequality in the distribution of world resources. It attempts to reduce these inequalities, but it is not at all clear what type of a structure NIEO visualizes. All the same, it provides a basis for the North-South dialogue and it has already become a base for negotiations, or confrontations, between North and South as witnessed at the Paris Conference in 1975. It has been interpreted as a charter of demands by the South to the North for transfer of real resources to it. Various principles determine how this transfer should take place, and which countries in the South should receive how much, assuming these are transferred.

In the previous parts of the book we have developed various analytical tools: division of the developing countries (south) into three groups (namely high income, high technology, and others) and CDS and

ADS, or the division of a country into two societies, rich (R) and poor (P). Accordingly, we shall analyze the issue involved in NIEO by these tools.

DEVELOPED COUNTRIES AND HIGH INCOME DEVELOPING COUNTRIES (HIDC)

As pointed out in Part I, high income developing countries (HIDC) are, basically, petroleum-exporting Arab countries. As a result of an increase in the price of oil, in the last few years they have received so much foreign exchange in so short a time that they have not had the time and capacity to analyze, appreciate, and understand the enormous implications of such large resources falling into their hands. At present the elites of these countries are satisfying, literally, their whimsies. However, a few issues are becoming clear. One, the major market for petroleum is in the developed economies. The affluence of HIDC, thus, depends upon growth in the North. This is a case of genuine interdependence — the level of dependence of one on the other is more or less equal — and it implies cooperation between HIDC and the North and not a confrontation.

Two, the increase in the price of gasoline has started a dynamic process to reach the new equilibrium. There is going to be a shift in the demand curve resulting from such measures as conservation and the switch from petroleum to other energy sources.(14) Thus the composition of production will undergo a major change. Already there is a move towards the production of gasoline-efficient automobiles in the United States. Similarly, the increase in price will encourage increases in the supply of petroleum and other substitutes. In view of the unusual time lag involved and the nature of technological change, it is not clear if, when, and where such a new equilibrium will take place. Much depends upon the dynamics of these processes. The nature of these processes have a major effect on the relationship between the North and HIDC. These processes can lead to both genuine cooperation and confrontation. At this stage it is not clear which direction these processes will take.

Three, having obtained large amounts of foreign exchange, the HIDC have to maintain the purchasing power of these quantities. This poses questions of investment and international monetary systems. It is clear that the HIDC have not worked out these problems. For the time being these sums are being invested in, and spent on goods of, the developed countries. This leads to further reasons for cooperation between HIDC and the North. Four, the petroleum is a nonrenewable source. Eventually, it will run out.(15) There is an obvious need to conserve these resources and to diversify the economies. The immediate effect has been an increase in imports of all sorts of commodities. Arab countries are now emphasizing the import of technology from developed and some developing countries. Technology, particularly high technology, involves all sorts of ancillary inputs and linkages. There has been an enormous increase in the import of skilled and unskilled labor from South Asian countries into HIDC. Technology, thus, brings further cooperation between HDIC and the North.

In view of all this, it is not clear how NIEO will affect the relationships between HIDC and the North. None of the principles on which NIEO is expected to be based refer to the interests of HIDC.(16)

DEVELOPED COUNTRIES AND HIGH TECHNOLOGY DEVELOPING COUNTRIES (HTDC)

There are 8 to 15 developing countries with developed industrial structure employing levels of science and technology. These have been called High Technology Developing Countries (HTDC), for want of a better term.(17) These countries are: Argentina, Brazil, Chile, Egypt, India, Indonesia, Iran, Kenya, Korea, Mexico, Nigeria, Pakistan, Peru, Philippines, and Syria. They obviously produce manufactured goods. The populations in these countries are large and their geographical size is also huge. These are also the countries which have moved toward dual societies of rich (R) and poor (P). Interestingly enough, virtually all of these countries, except Egypt, India and Mexico, have military dictatorships and are police states.(18)

The major interests of the governments of HTDC may be defined in terms of maintenance of, and growth in, the industrial sector; maintenance of military establishments; and provision of minimal food for the populations in urban areas. The industrial sectors in these countries is highly sophisticated, producing a range of manufactured goods. However, the domestic market for the absorption of these manufactured goods is limited. This market places a heavy constraint on the existence and growth of the industrial sector. The imperatives of this sector are foreign markets, and the most important markets are in the developed countries. It is this need, and interest, of HTDC that has been reflected in the language and clauses of NIEO, in terms of access to the markets of developed countries.(19) With these goods, manufacturers of the North compete directly, and indirectly, with HTDC. In many developed countries, such imports are considered harmful to employment opportunities. There is, thus, a direct conflict between the North and HTDC arising out of the imperatives of the industrial sector of HTDC. Not only is there a conflict, but the bargaining strength of HTDC is limited, at best. There is no interdependence with this system of export and, on the one hand, these exports put HTDC in a position of "dependency."

Exports are not the only imperative of the industrial sector of HTDC. There is another imperative from the import side. The industrial sector requires imports of technology, and it is thus no accident that transfer of technology from the North to HTDC – and other developing countries – has been given an important place in the NIEO. In the program of action, it is suggested that all efforts should be made:

a) To formulate an international code of conduct for the transfer of technology corresponding to needs and conditions prevalent in developing countries;

b) To give access on improved terms to modern technology and the adaptation of that technology, as appropriate, to specific economic, social and ecological conditions and varying stages of development in developing countries;

c) To expend significantly assistance from developed to developing countries in programmes of research and development and creation of suitable indigenous technology;

d) To adapt commercial practices governing transfer of technology to the requirements of the developing countries, and to prevent abuse of the rights of sellers;

e) To promote international cooperation in research and development in exploration and exploitation, conservation and legitimate utilization of national resources and all sources of energy.(20) (Emphasis added)

Since major R&D is conducted in the North, HTDC are net importers. In this case also, the relationship between the North and HTDC is not an interrelationship, but one of dependency of HTDC on the North.(21)

The other objectives regarding military establishment also force a "dependency" of HTDC on the North.(22) It is the North which is the largest supplier of armaments; the south is a buyer. The HTDC buy the more sophisticated parts of military equipment and they do not have much to offer to the North. The strange thing is that the HTDC, in spite of their established industrial sector, cannot produce enough food for their own populations. They are regularly in the market for food.(23)

The overall relationship between the North and HTDC is thus one of "dependency." NIEO, therefore, is of great importance to HTDC; but would NIEO eliminate this dependency?

DEVELOPED COUNTRIES
AND OTHER DEVELOPED COUNTRIES (ODC)

Other developing countries (ODC) refer to the bulk of developing countries, numbering anywhere from 120-140. There is a large variation among them and they are, by and large, small. The per capita incomes in these countries vary. The use and level of science and technology in these countries is limited; some of them do not have any industrial sector. These are in various stages of development; some are fully self-sufficient in terms of their needs, others have to import even their basics. Their exports are, basically, individual raw materials or individual crops.

The relationship between developed countries and ODC range between interdependence – like the interdependence between the North and HIDC – and dependence – like the dependence of HTDC on the North. The extent of interdependence follows from the nature of imports and exports of a particular member of ODC. If the export is a mineral raw material which is an essential input in the production structure of the North, the relationship has the potential of interdependence. This interdependence, however, may not exist in view of the smallness of the particular member of ODC. It is in view of such potentiality of interdependence that there is talk of forming groups such as OPEC. Similarly, if the particular country is able to produce and satisfy its basic needs, it does not depend upon imports from the North.(24)

On the other hand, if the export of a particular country is a commodity or crop which is not essential to the production structure of the North, it does not generate the potential for an interdependence. The wide fluctuations experienced in the export earnings of developing countries follows from such exports. In case export earnings are needed for development purposes, these fluctuations generate a relationship of "dependence." Similarly, if the development plans and policies of a country depend upon imports from the North, these imports also generate dependence.

NIEO deals with some of these issues. It seeks stabilization of export earnings, transfer of financial resources, stabilization and increase in the prices of raw materials, concessionary imports from the North, and so forth. In other terms, NIEO conceptualizes some of the problems of ODC and suggests some possible solutions.

NORTH-SOUTH DIALOGUE – CDS OR ADS? OR NIEO AND BASIC NEEDS

The world produces and uses a certain amount of resources every year. These resources are distributed in, and used by, various countries. In the world today, some countries, particularly developed countries in the North, have managed to obtain a share of resources far greater than can be justified either by their contribution or by needs. Other countries are left with a much smaller share. For example, the United States alone consumes 40 percent of the world's resources. There is, therefore, an unequal distribution of world resources among countries, and it has been going on for a number of decades. The inequality of this distribution has become an essential feature of the international economic order and it is prepetuated by an international pricing system.(25) The unequal international distribution of resources, as well as the international price system, is maintained and furthered by military strength and technology.(26) NIEO is based on a recognition of the facts of inequality in the distribution of resources, and the bias in the price system. It attempts to reduce the inequalities, and redress the bias. Even if one assumes that NIEO is successful and becomes a reality, the question is: Can it do so? Our analysis suggests that NIEO cannot do so; it cannot change the process of distribution of world resources towards a more "equitable" distribution. Furthermore, it cannot remove the biases in the international price system. The reasons are as follows.

The process of inequitable distribution of world resources over time has created an international structure based on inequality. NIEO does not address this structure. It addresses the annual increments in this structure, not the structure itself. NIEO, at best, defines a base for negotiations between North and South. Since the whole declaration deals with the transfer from the North to South, it is based on the assumption that the North is the one to give, the South the one to take. Thus, at best NIEO is a strategy to seek "concessions" from the North. If NIEO becomes a reality, if the policies declared in NIEO are successful, it will maintain the existing unequal structure.(27) The only change will be that some of the developing countries will join with the developed countries. That is, it will be successful in changing the membership in North and South, but not in eliminating the differences.

The reason why NIEO cannot change the existing unequal international economic structure is that it does not deal with two basic elements in the maintenance of, and bias in, the international price system, namely military strength and technology. Interestingly, there is no reference in the NIEO to unequal military strength and unduly large military expenditures by both North and South. Economic structure is not independent of military strength. The concept of technology implied in NIEO is that of "large" and "hard" technology – now prevalent in the North. NIEO thus deals with issues of "technology transfers." As we have already pointed out in Part II on CDS, the implications of technology transfers is a CDS. NIEO thus implies and is based on CDS. As we have suggested in the section above, technology for development of the industrial structure is one of the sources of creating, maintaining, and accentuating dependency of the South and North (such as the dependency of HTDC on the North). NIEO is based on an illusion of "catching up" since it misses completely the experience of increased dependence of countries that have tried to do so.

NIEO attempts to make minor changes in the international price system by seeking "concessions" such as stabilization of commodity prices, access to markets in the North, and changes in the voting powers in the international institutions. It does not and cannot eliminate the biases in the international price system. NIEO is based on the assumption that the international price system is independent of domestic price systems in various countries. This assumption is fundamentally wrong.(28) In actual fact, the international price system is highly interrelated with a domestic price system; the international price system operates through a domestic price system. In view of this interrelationship, the policies implied by NIEO in seeking concessions further accentuate the existing biases in the international price system. Thus commodity price stabilization, access to the markets in the North, and transfer of technology from the North emphasize domestic policies in favor of export promotion. These policies distort domestic prices which in turn maintain the biases in the international price system. More generally, NIEO is based on CDS. As we have argued in Part II, CDS is not a solution to the problems of international inequality. It is a cause of it.

There is a serious implication of NIEO in terms of basic human rights. If NIEO is effective in putting in place an industrial sector in a developing country via technology transfers and other help from the

North, the logic of this sector is to further the movement towards a dual society of rich (R) and poor (P). This leads to the establishment of a police state. As we pointed out, it is not a pure accident that virtually all HTDC contain large elements of a police state. They are police states.(29) In that case, NIEO is counterproductive. Propaganda not withstanding, police states are most inefficient in terms of development, however development is defined. The inefficiency arises from a continuous tension and the impossibility of defining priorities intelligently.   The inefficiencies of a police state are matched by their inhumanities. NIEO, thus, may have been misconceived.

The issue of the gross inequalities in the international economic – and price – structure is a serious one. It generates tensions and is a source of future wars. For peace, and happiness, there is a need for an equitable international order. It requires a fundamental change in the structure, not independent of changes within the internal structure of both North and South. Just as the South has problems of poverty, unemployment, inequalities and underdevelopment, the North also has problems of stagflation, unemployment, inequalities, and overdevelopment. Just as the South needs stable export prices and market access in the North, the North also needs raw materials and economic growth. They both have a mutual interest in an alternative international structure.

The existing international structure is based on CDS. ADS provides a base for an alternative international order and a dialogue between all nations. The alternative structure should shift, adopt, and accentuate favorable existing elements. In ADS the philosophy of development is a different one; it involves development of every human being, in both material and nonmaterial dimensions. It is a long-term strategy concentrating on ends and appropriate means. An international formulation of ADS would have some of the following implications:

1) The international order should provide an opportunity for every nation to develop on the basis of its own needs and resources. In other words, development is development and not modernization or westernization. Trade in such a process will reflect mutual needs and interdependence, not dependence of one on the other. Such development implies growth in the world economy.

2) The international community can plan how to best utilize the world's resources. Such planning is possible only if there is genuine participation. In other words, the international order has to be participatory and not based on military strength.(30) It will also need international institutions that are not hierarchical like the existing institutions. Such institutions could be networks with open memberships.

3) The technology involved will be appropriate technology, as we have discussed at length. Appropriate technology provides a base for equality; the existing technology immediately forces an inequality.

4) Such an order will involve free flows of goods and people. Existing structure puts all sorts of constraints on such flows.

5) Instead of the dialogue between North and South being a negotiation for concessions and/or confrontation, such an order will elicit cooperation.

SOME POLICY ISSUES

On the basis of our analysis we can suggest some policy issues for a North-South dialogue. It needs to be emphasized that this is only a tentative list. It is neither comprehensive nor exhaustive, nor can it be so.

1) Common Problems. It is our contention that both North and South are facing problems arising from the CDS paths and investments in "high/hard" technologies. Both the North and South are facing problems of persistent trade deficits, unemployment, and inequalities; both are finding that military expenditures are becoming too large and excessive a burden; the North is facing stagflation and environmental degradation. It finds there is less scope for maneuvering. The South finds itself burdened by poverty. The international economic and price structure, thus, is creating problems for both. There is a need to debate and discuss these common issues; the South should get interested in the problems of the North.

2) Planning. Give world interdependence, there is a need to formulate a planning process in order to utilize the world's resources efficiently. The North needs raw materials, the South needs food and technology. South must recognize and understand the problems the North faces. It should join in the discussions of these problems. Also, a cooperative effort should develop to make changes in the international structure and price mechanism consistent with the desired changes in the internal structure within the countries of North and South. Such discussions and cooperative efforts will generate an atmosphere of mutual trust instead of the existing climate of distrust. Such effort involves determination of people's needs; determination of world resources; and technologies and mechanisms of allocation of these resources to satisfy the needs. Information on some of these issues is collected, haphazardly, by international agencies such as the World Bank. However, there is a need for a more effective body based on principles of equality.

3) AT. As we have argued, a majority of the problems in countries both of North and South are amenable to solutions

via AT. Since AT builds on existing traditions, it provides an excellent base for cooperation between North and South. In this area, all countries are equal. Both North and South can learn from each other. There is thus a need to develop ADS institutions for the exchange of information, people, and research and development associated with AT.

4) <u>Nongovernmental Organizations (NGO)</u>. ADS speaks to long-term issues and involves changes in internal as well as international structures. Government bureaucracies and business interests, unfortunately, are much too concerned with short-term interests and existing structures for various reasons of national prestige, pressure groups, vested interests, profits, and so forth. NGOs, on the other hand, can be and generally are motivated by long-term interests and ideals. Being by necessity part of the grass roots and involved with people, these NGOs have a very important role to play.

16 Cooperation Between Developing Countries

INTRODUCTION

The Declaration and Program of Action on the Establishment of a New International Economic Order contains the following item on the promotion of cooperation among developing countries.

1. Collective self-reliance and growing co-operation among developing countries will further strengthen their role in the new international economic order. Developing countries, with a view to expanding co-operation at the regional, subregional and interregional levels should take further steps, inter alia:

a) To support the establishment and/or improvement of appropriate mechanisms to defend the prices of their exportable commodities and to improve access to and to stabilize markets for them. In this context the increasingly effective mobilization by the whole group of oil exporting countries of their natural resources for the benefit of their economic development is to be welcomed. At the same time there is the paramount need for co-operation among the developing countries in evolving urgently and in a spirit of solidarity all possible means to assist developing countries to cope with the immediate problems resulting from this legitimate and perfectly justified action. The measures already taken in this regard are a positive indication of the evolving co-operation between developing countries.

b) To protect their inalienable right to permanent sovereignty over their natural resources.

c) To promote, establish or strengthen economic integration at the regional and subregional levels.

d) To increase considerably their imports from other developing countries.

e) No developing country should accord to imports from developed countries more favorable treatment than that accorded to imports from developing countries. Taking into account the existing international agreements, current limitations and possibilities and also their future evolution, preferential treatment should be given to the procurement of import requirements from other developing countries. Wherever possible, preferential treatment should be given to imports from developing countries and the exports of those countries.

f) To promote close co-operation in the fields of finance, credit relations and monetary issues, including the development of credit relations on a preferential basis and on favourable terms.

g) To strengthen efforts which are already being made by developing countries to utilize available financial resources for financing development in the developing countries through investment, financing of export-oriented and emergency projects and other long-term assistance.

h) To promote and establish effective instruments of co-operation in the fields of industry, science and technology, transport, shipping and mass communication media.

2. Developed countries should support initiatives in the regional, subregional and interregional co-operation of developing countries through the extension of financial and technical assistance through more effective and concrete actions, particularly in the field of commercial policy.(1)

Since the special meeting of the General Assembly, a number of conferences have taken place on specifically this issue. The latest took place in Buenos Aires, from August 30 to September 12, 1978.(2)

To analyze the various issues of technical cooperation, let us examine the interests and relationship between the three groups of developing coutnries.

HIGH INCOME DEVELOPING COUNTRIES (HIDC)
AND HIGH TECHNOLOGY DEVELOPING COUNTRIES (HTDC)

HTDC have an industrial sector that needs inputs from abroad and produces goods for export to foreign markets. One of the inputs they need is petroleum. They also need foreign exchange for the importation of technology to maintain and update their industrial structure. They thus need both foreign markets and long-term loans in foreign exchange.

HIDC provide all these facilities; they have oil to export, large surpluses of foreign exchange, and a rich internal market for the manufactured goods. HTDC, then, find the relationships with HIDC very satisfactory. HIDC, on their part, are in the process of expending newly found large resources in foreign exchange. They are involved in importing high technology commodities and complexes, such as universities, and hospitals.   They are trying, hopefully, to diversify their services and consumption. In this process, they need unskilled labor for service, skilled labor for mechanized processes, and professionals in terms of engineers and scientists. They are also interested in as diverse a consumption level as possible. In their affluence, there is scope for every country to provide a service. Labor is particularly available from some of the HTDC, particularly those in South Asia. There is thus a potential for meaningful cooperation between HIDC and some of the HTDC.

However, the opportunity for cooperation is not without potential for conflict. This conflict arises both from HTDC and HIDC. Let us take the HTDC first. The affluence of HIDC depends upon a continuous increase in the price of petroleum; and its rise is determined by demand in the market of the North. However, HTDC also have to buy the petroleum at international prices. The increase in petroleum prices, thus, raises the cost structure in the industrial sector of HTDC. This cost structure is already quite high in view of higher capital/output ratios and unused capacities. If HTDC are not able to reduce or maintain their existing cost structure, they face difficulties in the export of goods. The HTDC, therefore, have to find substitutes for petroleum as an input. The extent of substitution and the increases in petroleum price raises the potential for conflict.(3)

The existing policies in HIDC involve large imports of technological goods and systems on the one hand, and labor on the other.(4) The technology consumption requires a large number of people with technological skills, but these skills for the moment are nonexistent. They are being provided by imported labor. The consumers are local Arabs while the skilled labor is Egyptian, Indian, Palestinian, and Pakistani. The differences in races always pose a racial situation.(5) In view of the dependence of Arabs on the skills of imported labor, this situation can be a source of potential conflict. This is true for the unskilled labor also. Furthermore, in some countries like Kuwait, the imported labor forms a sizeable proportion of the total population. The conflict arises whenever there are any type of disturbances in the coutnry – economic or political.(6)

HIGH INCOME DEVELOPING COUNTRIES (HIDC)
AND OTHER DEVELOPING COUNTRIES (ODC)

In view of the facts that HIDC have acquired large wealth and hence international economic influence only recently, and that these are small countries with small populations, their relations with a large number of countries in ODC are at best limited. At present HIDC can offer ODC two things, petroleum, and foreign exchange. However, HIDC need little from ODC, except from particular countries. Their only needs then are defined by their international politics. In other words, HIDC have certian political interests. They do need the support of ODC in these interests, i.e. votes in the UN. On their part, ODC have much to gain from their relations with HIDC. Even if ODC do not need petroleum, they do need foreign exchange and markets where they can obtain it.

In case the HIDC and ODC follow sensible policies, there is a potential for cooperation between ODC and HIDC in the future. These policies relate to the nature of development in the ODC. If ODC follow policies of ADS, their economies will grow. Furthermore, the ODC will have many things to offer HIDC. Even the application of "appropriate technologies" – as discussed in Chapter 10 – will increase the demand for petroleum, by increases in total output. This will thus diversify demand for petroleum and reduce HIDC dependence on the markets in the North. To develop their own potential, via ADS, ODC will need loans in foreign exchange at reduced rates and for long terms. (This is something HIDC can provide.) On the other hand, if ODC follow the CDS path, ODC will have nothing to offer to HIDC even though they will become dependent upon the goods and foreign exchange from HIDC.

HIGH TECHNOLOGY DEVELOPING COUNTRIES (HTDC) AND OTHER DEVELOPING COUNTRIES (ODC)

At present HTDC desperately need markets for their exports of manufactured goods and foreign exchange for the import of technology and basic goods such as food. Both are necessary to maintain their industrial sector. The ODC can provide markets, but they do not have the foreign exchange to pay for it. The needs of ODC, at present, are technology, foreign exchange, and markets for their exports. Their exports are made up of two different types of products: crops such as bananas, cocoa, coffee, sugar, and tea which are basically consumer good crops; and raw materials such as bauxite, and copper. The HTDC do not, and cannot, provide markets for their crops because these countries themselves produce these crops. Also, these crops can be exported only when they are processed and converted into industrial goods. The cost of industrial goods is higher and the HTDC do not have markets for industrial goods.(7) HTDC do not provide markets for the raw materials of ODC because the level of industrialization in HTDC has only reached the point that they can use their own raw materials.(8) HTDC themselves need foreign exchange and cannot help ODC. The only real need HTDC can satisfy is in the area of technology. It is because of these reasons that the trade between HTDC and ODC has remained relatively small.

The technology that HTDC can provide is the technology that HTDC have themselves imported from the countries in the North. The only advantage of technology from HTDC is that it is cheaper, particularly the software component of such technology. One would also hope that HTDC have learned a little in the process of establishing, and being saddled with, their industrial sector.(9) There are two problems here. It is not easy to define a market for technology. As we have argued, technology is a process. Particular components may have a market. even in the case of well-defined machinery components, the market is not obvious because these components are, at best, similar. They are never the same. To determine what prices of which components are cheaper, then, becomes all the more difficult. Many of the problems of choice of technology enter here.(10)

The ODC also want – just like HTDC and for the same reasons – the most up-to-date and modern technology. The HTDC do not, and cannot, offer the most modern technology. The members of ODC have to take the idea of buying technology from HTDC seriously, otherwise these factors work against this process.(11) It is for this reason that HTDC are particularly interested in the clause "to increase considerably their imports from other developing countries" in the Program of Action.(12)

The transfer of technology from HTDC to ODC has a large potential for conflict if ODC follow the CDS path and assume that the HTDC have moved ahead in this path. In other words, if ODC consider themselves in the initial stages of development and HTDC in the higher stage of development and technology, this will imply and create a hierarchy of development and of the level and sophistication of technology. In this hierarchy, countries in the North are at the top, ODC at the bottom, and HTDC in between. HTDC in this case will be junior partners of the North in transferring inappropriate technology to ODC. We have already commented on the relationship between the North and HTDC. This relationship is one of dependence of HTDC on the North.(13) Such a technological transfer will then create a "dependence hierarchy"; ODC will be dependent on HTDC and HTDC on the North.(14) The reasons for dependency arise from the inappropriateness of technology transfers. If HTDC, in spite of their developed technology sector, find it difficult to make rational technological choices, the task for ODC to make such choices is all the more difficult. Since all HTDC have to export is "inappropriate" technology, there is no way such technology becomes "appropriate" as it reaches ODC – except by a miracle or an accident.

This is not to say that there is no potential for cooperation between HTDC and ODC. There is. Much of this cooperation depends on, and ties in, the process towards regional integration. Integration does not and cannot take place unless it is based on principles and institutions of equity and equality. In this context advantages lie in the ADS and AT.

TECHNICAL COOPERATION, ADS AND AT

Genuine cooperation is based on equality and interdependence. Interdependence may be defined as equal dependence. At present there are two sources of inequality among the developing countries, namely military strength and technology. For the past 30 years various developing coutries have been acquiring military muscle.(15) A number of countries, particularly in the HTDC group, have become quite strong militarily, such as Argentina, Brazil, India, and Iran. There are now strong military countries in every region. Just as in a more equitable world the developed countries have to reduce significantly the size of their armaments, similarly for meaningful cooperation these countries have to start reducing their military acquisition and hardware. And similarly, some countries have larger potential in terms of technological sophistication. In the last section we have already commented on the areas of interest and conflict as a result of this technological gap. Furthermore, many developing countries have dual societies of rich and poor; they create large internal political instabilities and a continuous threat of disorder. Such divisions are inimical to cooperation among countries, particularly among neighbors.(16) Meaningful cooperation among developing countries will, thus, involve a process for the resolution of some of the domestic problems, particularly those dealing with poverty, unemployment, and inequalities of economic and political power. The weakness of the Program of Action regarding NIEO is that it completely ignores this vital, even necessary, condition for cooperation. As we have argued in Part III, ADS addresses these very problems and it thus provides a sound base for the formulation of genuine cooperation between developing countries.

Science and technology associated with ADS is Appropriate Technology (AT). We have derived AT from the objectives of ADS.(17) It is thus a part of the solution of the above mentioned problems. AT provides an excellent base for technical cooperation among developing countries with two clear advantages. It relates directly to the problems of basic needs of the people. It reduces the differences between rich and poor and provides an integrative force within the country. As a result, it contributes to the process of reducing internal tensions and disorders and thereby the need for military strengths. On the other hand, it develops capacities and capabilities of large numbers of people to find scientific solutions to the social, economic, and political problems. Since AT is based on and derived from traditions, every region, every country, and every area has a particular AT, or ADS, to contribute to the common pool among developing countries. No particular country has an advantage in terms of AT. There is thus no relationship of "dependence." In view of the existence of a pool of ATs in various stages of development and sophistication, there is a basis for important and fruitful cooperation among the developing countries. Such cooperation would deal with issues of appreciation and recognition of a particular AT; possibilities of its adaptation and exchange; and the need for designing, research, and development. The cooperative efforts will involve clarification, comprehension, debate, and discussion about these issues regarding AT.

Technologies, by themselves, are not complete solutions. These operate within social, political, and economic institutions. The success of ADS, AT, and the meaningful cooperation between developing countries require the existence of responsive institutions. We have already discussed elements of these institutions.(18) Like AT, these institutions also have the same two advantages. Derived from ADS, they address the fundamental problems of inequality, poverty, and unemployment. They promote the process that solves these problems. Thereby, these are instrumental in reducing internal tensions and disorder. On the other hand, they encourage integrative forces in the society. These institutions are derived from the traditions and cultures of the societies. There are at present no single best institutions. Nor can there be. Accordingly, every country, region, and area has a contribution to make in this respect. Here again all countries, big and small, start from equality. Every country can learn from the institutions in the other. Here, too, there is a pool of institutions in various stages of development which provides another important and fruitful basis for cooperation among the developing countries. The cooperative effort will concentrate on such issues as examination, analysis and revitalization of existing responsive institutions; formulation, creation, and development of other responsive institutions; and adaptation of revitalized and newly developed institutions.

The Program of Action regarding NIEO suggests cooperation among the developing countries in the form of OPEC. Such cooperation will be useful, particularly for bargaining with the North. However, it is, by its very nature, short term.(19) The issues we have raised are long-term issues. NIEO also suggests setting up of regional units. However, it is not clear what the nature and objectives of such institutions will be. It emphasizes national research and training centers of multinational scope.(20) This deals with strengthening the existing institutions without considering if they are appropriate or inappropriate.(21) Furthermore, by building on what exists, it compounds inequality. However, in the long run, cooperative efforts among developing countries, based on AT and ADS institutions, when successful must naturally take the form of regional integration. This is so because of the fact that the formation of many developing countries in the various regions of the world is the result of colonial processes. Cooperation based on AT and ADS is the ultimate removal of remnants of all colonial influences.(22)

SOME POLICY ISSUES

On the basis of our analysis, we suggest the following issues for cooperation between developing countries. This is, however, a suggestive and tentative list; it is by no means comprehensive or exhaustive. (The cooperating countries may set up bodies relevant to the tasks, based on democratic principles.)

1) Monitoring of ADS Objectives. Such bodies will monitor the progress towards the objectives of ADS and reductions in military expenditures and armies; in level and magnitude of poverty; in unemployment and underemployment; in political and economic inequalities; in potentials for internal disorders and chaos, and so forth. The existing UN regional bodies, like ECLA, ECAFE, and ECA, can be employed to perform these functions.(23)

2) <u>Establishment of Monetary Zones</u>. The formation of money zones can provide collective self-reliance and strength to the countries in the zone. Such zones can take care of the problems regarding stabilization of commodity prices and planning of production and distribution of food from surplus to deficit countries. Countries in the HIDC group, obviously, have little resources to produce their own food.

3) <u>AT</u>. Such cooperative bodies will shift ATs in various parts of different countries. This involves appreciation and recognition of ATs embedded in traditions and cultures. Having sifted some ATs, these bodies will study the possibilities of their adaptation to the existing situation and means of exchanging them with ATs in other countries. Some of these bodies will design, research, and develop ATs.

4) <u>ADS Institutions</u>. Such cooperative efforts will involve analysis and examinations of existing institutions with a view to revitalization. There is also a need to formulate, create, and develop new institutions responsive to the objectives of ADS.(24) Some bodies will look into the possibilities of adaptation of these institutions for use in other countries.

5) <u>Role of Nongovernment Bodies, i.e., People</u>. In this cooperative effort, real people have the most important part to play. It is they who define the needs, the objectives of ADS, the appropriateness of technology, and so forth. Since the cost of such cooperative efforts is low, it does not require fancy establishments and hierarchies for people to get involved. The bodies implied in these activities can be as simple as networks. The nongovernment bodies, thus, have particularly meaningful contributions to make in this effort.

17 For an Equitable World Order

The existing international economic and technological structure is based on an unequal distribution of world resources among various countries. Similarly, the political and economic system within countries is based on inequalities. There has been, however, a very slow reduction in these inequalities over the past few decades. The current neocolonial situation is certainly an improvement of the colonial situation of only two or three decades ago. In Rhodesia and South Africa this colonialism still lingers. These inequalities are maintained, both internationally and nationally, by a military with police power, a price system, and "high" technology. In academic debates these three areas are separated, as if independent of each other. In actual fact these are highly interrelated. They feed on each other and have cumulative effects. Thus internationally powerful countries with large military strength (United States) can, and do, prop up military and minority dictatorships (Iran) in some countries. These minority dictatorships buy high technology that gives them control of production and people in the country. These technologies, in their turn, define and distort relative prices in favor of high technology products. These technologies also produce or make available goods — particularly raw materials (petroleum) — for the markets of the powerful countries at prices favorable to their markets. This transfer of resources from the propped up countries provides resources for further strengthening the military strength of the powerful country. Since military strength, biased price system, and high technology form a circle, one can start at any other point, such as prices or technology, and get back to the same point.

Prima facie, it seems that the countries with military strength, price systems biased in their favor, and high technology have all the advantages. These advantages, however, are superficial and strictly in the short run. Let us take the case of a country with military strength, a price structure favorable to it, and a high technology within. The United States is such a country. The military strength imposes a burden; after Vietnam, the United States has realized it cannot afford both guns

161

and butter. Since Vietnam, and because of it, its standard of living has started sliding downwards. Again, biased prices distort the production structure. Thus, the low price of gasoline has made the U.S. dependent upon the imports of oil. Increase in the price of oil is causing all sorts of painful adjustments. The size of automobiles is decreasing; the electric utilities are in a squeeze; so is the consumer. The high technology, in its turn, pushes costs up. It forces the problems of stagflation, environment pollution, and so forth. No wonder there are a number of studies arguing for "soft" technology paths. It is difficult to keep the minority/military dictatorship propped up for long. The former Iranian government, in its last days, caused all sorts of nervous tension in the United States government. When such propped up governments fall, the cost to the economy and the country is high. The country whose government is propped up suffers choas, internal disorders, and even civil war.

Similarly, the ruling minority elites seem to have all the advantages in a country that borders on a police state, imports high technology, and maintains a distorted and biased price system favorable to the goods demanded by these elites. These elites do have the advantages, particularly in terms of privileges. However, these privileges and advantages are also superficial. These privileges have costs, particularly in the long term. The elites who make use of these privileges have to pay the price, even though in the long run. These governments always face the threat of internal disorders. They are dependent upon foreign governments and interest groups who can drop them at any time. The life they lead is full of insecurity. Even materially there are continuous failures of technology, because of work stoppages, strikes, and even riots. If the elite groups are civilian, they are always threatened by military take-over that is always in the background. If the elites are military dictators, they are always worried about internal factions. Legitimacy remains elusive. On a historical time scale, just as colonized people suffered colonialism for centuries and then rose up so that even large empires crumbled in a matter of two decades, the neocolonial system is also subject to fast decay. There is evidence that this process of decay has started.

It is therefore in the self-interests of the governments and elites of both developed and developing countries to work cooperatively towards the formulation of an alternative international order which reduces and eventually eliminates inequalities, armaments, biases in the price system, and technological inappropriateness, both nationally and inter-nationally. Such an equitable world order is not only in the interests of the elites; it is appropriate to the imperatives of economic efficiency in view of the new cost structure reflecting changing world resources supplies. In addition, it is ecologically viable, ethically decent, and morally desirable.

Since science and technology are an important element in the maintenance and growth of an unequal national and international structure, there is a need to rethink science and technology. In this context, science and technology may be considered a matter of access. The issue of "access" can be looked at from two angles: from the point of view of control, and on the basis of impact. From the idea of access, two separate science and technology forms can be derived. Science and technology forms that are controlled by few and have impacts on many are one group. The nuclear power plant is an example of such a form. A few operators of a nuclear power plant can affect the lives of hundreds and thousands of electricity users. Generally, "high" and "hard" technologies lie in this category.

The extreme case of such science and technology is when a very small number of persons can affect the lives of millions. Atom bombs provide such an example. In the past few decades, military science and technology have moved in this direction. Since a small number of people can through such science and technology affect a much larger number of people, such science and technology provide a base for "privilege." Because of this privilege science and technology lead to inequalities. Furthermore, such science and technologies are large and involve huge outlays, thereby denying access to the majority of people.

Generally speaking, the control of any science and technology is in the hands of a small number of persons. If the impact is also on a small number of persons, we obtain the second group of science and technologies. An example of such science and technology is the use of a bicycle for transportation. By such science and technology few people cannot affect the activities of a large number of people. Such science and technology are generally small and involve small total outlays; they are thus accessible to a large number of people and they allow people to define their own objectives and solve their own problems. These are the science and technology associated with ADS and are often called "soft" "appropriate" technologies. They reduce inequalities among people.

In an ideal world, there will be only soft technologies. Such an ideal world does not exist; it is hundreds of years away. The reality, instead, is made up of a mixture of hard and appropriate technologies. This mixture varies with the country. In the past few decades policies have emphasized hard science and technology paths. But over time such science and technology have developed a momentum of their own which has now created its own side effects and externalities, and has halted the momentum of hard technologies. One result of these side effects has been to encourage a trend towards appropriate science and technology. Currently, the world is in this transition stage. The hard science and technology have reached a threshold level. It will be fool hardy to pretend that these do not exist or that they will disappear in the foreseeable future. Many of these technologies can be used and adapted in the movement toward an equitable world order. On the other hand, the emphasis on appropriate technologies is still recent. These technologies have not yet developed their own momentum. This momentum, by its very nature, requires a widespread use of appropriate science and technologies and it is essential for the movement towards an equitable world order.

The logic and imperatives of history are clear. The world can survive if, and only if, sufficient progress can be made toward an equitable world order. Such progress, in its turn, provides imperatives for science and technology policies. These policies need to emphasize and facilitate the adoption of appropriate technologies and they must also adapt hard technologies to the ADS objectives of "development of <u>every</u> human being." This book has dealt with these issues. Hopefully, it is a step towards an equitable world order.

Appendix

Science, Technology and Per Capita Income for Countries in the World

Countries	Per Capita GNP U.S. $ 1974	Scientists and Engineers (No.)	Establishments- Statistical Units (No.)	Employment in Manufacturing (1,000)	Enrollments in Colleges and Universities (No.)
Low Income Countries					
Afghanistan	110	4,823 (66)	157 (74-75)		9,399 (73)
Bangladesh	100	23,500 (73-74)	1,417 (74-75)	293.2 (74)	183,833 (74)
Benin (Dahomey)	120				1,900 (74)
Bhutan	70				
Bolivia	280	10,925 (67)	853 (73)	21.0 (73)	49,850 (75)
Botswana	290	786 (73)	75 (74-75)	3.4 (74-75)	289 (74)
Burma	100	18,500 (75)			56,310 (72)
Burundi	90		31 (74)	1.5 (74)	517 (74)
Central African Rep.	210		37 (74)	5.7 (74)	318 (74)
Chad	100			645.4 (71-72)	484 (74)
Comoro Is.	230				
Egypt	280	593,254 (73)	4,812 (71-72)	606.4 (70)	408,235 (74)
Equatorial Guinea	290				201
Ethiopia	100		442 (69)	54.8 (73)	6,474 (73)
Gambia	170		421 (73)		111

Science, Technology and Per Capita Income for Countries in the World

Countries	Per Capita GNP U.S. $ 1974	Scientists and Engineers (No.)	Establishments- Statistical Units (No.)	Employment in Manufacturing (1,000)	Enrollments in Colleges and Universities (No.)
Guinea	120		703 (74-75)	16.6 (74-75)	1974 (70)
Haiti	170		904 (75)	17.4 (75)	1,494 (67)
India	1401	187,500 (72)	62,833 (73-74)		2,230,225 (74)
Indonesia	170		26,454 (73)	1,011.4 (72)	239,803 (72)
Kenya	200	3,955 (72)	507 (74)		11,351 (74)
Khmer Rep.	70				9,988 (72)
Laos	70				875
Lesotho	140				577 (74)
Macao	270				
Malagasy Rep.	180				
Malawi	130	1,253 (67)	130 (73)	25.7 (73)	1,153 (74)
Maldive Is.	100				
Mali	80		36 (70)	10.4 (70)	2,445 (74)
Mauritania	290				
Nepal	100			174.2 (74)	21,760 (74)
Niger	120		56 (72)		357 (74)
Nigeria	290	19,885 (70-71)	1,057 (74)	169.1 (72)	23,228 (72)
Pakistan	130	111,000 (73-74)	3,549 (70-71)	427.4 (70-71)	111,826 (73)
Rwanda	80		71 (74)		1,023 (74)
Sierra Leone	190				1,646 (73)
Sikkim	90				

Science, Technology and Per Capita Income for Countries in the World

Countries	Per Capita GNP U.S. $ 1974	Scientists and Engineers (No.)	Establishments- Statistical Units (No.)	Employment in Manufacturing (1,000)	Enrollments in Colleges and Universities (No.)
Somalia	90		259 (73)	6.2 (73)	958 (71)
Sri Lanka	130	7,457 (72)	2,344 (74)	114.6 (74)	14,568 (74)
Sudan	230	13,792 (71-72)			22,204 (75)
Tanzania	160	4,080 (68-69)	503 (73)	63.4 (73)	3,064 (75)
Togo	250	461 (71)	28 (72)	2.5 (70)	2,353 (75)
Tongo	210				
Uganda	240		466 (71)		5,618 (74)
Upper Volta	90		10 (75)	1.5 (75)	756 (74)
Vietnam Dem. Rep	130				88,104 (72)
	170		624 (73)		57,574 (71)
Yemen Arab Rep.	180		559 (75)		952
Yemen People's Rep.	220				934 (74)
Zaire	150	5,900 (70)	388 (72)	57.7 (72)	21,021

Lower-Middle Income Countries

Albania	530				28,668 (71)
Cameroon	300	3,500 (70)	331 (71-72)	21.3 (71)	5,533 (73)
Cape Verde Is.	470				
China People's Rep.	300				820,000 (62)
Columbia	500		6,066 (74)	(74)	148,613 (74)

Science, Technology and Per Capita Income for Countries in the World

Countries	Per Capita GNP U.S. $ 1974 (No.)	Scientists and Engineers (No.)	Establishments-Statistical Units (No.)	Employment in Manufacturing (1,000)	Enrollments in Colleges and Universities (No.)
Congo	470		59 (73)	16.3 (73)	3007 (74)
Cuba	640				68,504
Dominican Rep.	650		1,169 (74)	139.4 (74)	37,538 (73)
Ecuador	500		1,255 (74)	65.5 (74)	57,677 (72)
El Salvador	410		832 (72)	41.5 (72)	26,069 (74)
Ghana	430	6,897 (70)	362 (72)	58.7 (72)	8,023 (74)
Grenada	330				
Guatemala	580	5,551 (70)	1,950 (74)	58.8 (73)	21,878
Guinea-Bissau	390				
Guyana	500		86 (75)	28.7 (75)	2,307
Honduras	340		656 (71)	28.1 (71)	4,847 (70)
Ivory Coast	460				6,148
Jordan	430	4,288 (73)	7,399 (74)	15.4 (74)	9,302 (74)
Korea Dem. Rep.	390		23,729 (72)	946.4 (72)	214,653 (71)
Korea Rep. of	480	369,193 (73)	22,832 (74)		297,219
Liberia	390			1,274.7 (74)	1,710 (73)
Malaysia	680	30,000 (73)	219 (70)	3.4 (70)	
Mauritius	580	554 (70)	3,747 (72)	22.1 (72)	1,321 (74)
Mongolia	620	1,908 (72)	230 (74)	53.7 (75)	8,900 (72)
Morocco	480		1,492 (71)	87.8 (71)	24,092 (74)

Science, Technology and Per Capita Income for Countries in the World

Countries	Per Capita GNP U.S. $ 1974	Scientists and Engineers (No.)	Establishments- Statistical Units (No.)	Employment in Manufacturing (1,000)	Enrollments in Colleges and Universities (No.)
Mozambique	340		1,438 (73)	99.5 (73)	2,621 (72)
Nicaragua	670				12,519 (72)
Papua New Guinea	470				1,032 (70)
Paraguay	510				12,212 (73)
Philippines	330	94,302 (67)	13,313 (74)	315 (74)	678,343 (72)
Rhodesia	520		1,240 (71)	126.7 (71)	
Sao Tome and Principe	470				
Senegal	330				7,502 (74)
Swaziland	390		39 (72-73)		1,086
Syria	560		34,755 (75)		64,094
Thailand	310	18,827 (73)	6,276 (71)	194.9 (70)	75,432
Tunisia	650		1,150 (74)	69.5 (74)	17,540 (72)
Western Samoa	300				129
Zambia	520	5,900	612	47.9	4,681

Upper Middle Income Countries

Algeria	710		2,261 (69)	104.8 (69)	30,070
Angola	710		1,470 (72)		2,660
Argentina	1,520	333,000			497,727
Barbados	1,200		153 (73)	8.4 (72)	1,417 (73)

Science, Technology and Per Capita Income for Countries in the World

Countries	Per Capita GNP U.S. $ 1974	Scientists and Engineers (No.)	Establishments-Statistical Units (No.)	Employment in Manufacturing (1,000)	Enrollments in Colleges and Universities (No.)
Brazil	920	541,328 (74)	67,086 (73)	2,635 (73)	954,674 (74)
Bulgaria	1,890	190,458 (73)	2,138 (73)	1,096.5 (73)	127,319 (74)
Chile	830		1,270 (74)	1,155.8 (75)	149,647 (73)
Costa Rica	840		771 (68)	27.0 (68)	32,928
Cyprus	1,320	4,650 (69)	3,754 (75)	17.0 (75)	611 (74)
Fiji	840	315 (69)	366 (73)	9.2 (73)	1,089
Gabon	1,960				986 (74)
Guadeloupe	1,240				376 (69)
Hong Kong	1,610	41,000 (70)	31,034 (75)	613.4 (73)	38,265
Iran	1,250	127,793	6,056 (73-74)	348.6 (73-74)	135,354
Iraq	1,160	4,365 (74)	1,272 (74)	122.3 (74)	78,784
Jamaica	1,190		1,158 (72)	49.0 (72)	8,413
Malta	1,220	1,751 (73)	1,768 (74)	22.7 (74)	2,158
Martinique	1,540				1,673 (69)
Mexico	1,090	565,601 (71)			453,015
Netherland Antilles	1,590				781,320
Oman	1,660				
Panama	1,000		638 (74)	27.1 (74)	24,292 (74)
Peru	740	84,923 (74)	7,612 (73)	250.1 (73)	179,916

Science, Technology and Per Capita Income for Countries in the World

Countries	Per Capita GNP U.S. $ 1974	Scientists and Engineers (No.)	Establishments- Statistical Units (No.)	Employment in Manufacturing (1,000)	Enrollments in Colleges and Universities (No.)
Portugal	1,630		14,530 (74)	593.2 (74)	59,845
Reunion	1,550				639 (69)
Romania	1,960	274,541 (71)	1,731 (75)	2,802 (75)	152,728 (74)
South Africa	1,210		12,671 (71-72)	1,125 (71-72)	98,577
Surinam	1,120				(64)
Taiwan	810				
Trindad and Tobago	1,680		224 (72)	53.3 (72)	2,381 (70)
Turkey	750		5,952 (74)	657.7 (74)	218,934
Uruguay	1,190	20,069 (71)	5,648 (71)	213.4 (71)	32,627
Venezuela	1,970		7,350 (74)	284.3 (74)	213,542
Yugoslavia	1,310	217,989 (73)	3,574 (74)	1,577 (74)	359,651

High Income Countries

Australia	5,330		36,862 (74-75)	1,315 (73-74)	252,972
Austria	4,410	118,294 (72)	7,835 (74)	732.1 (74)	84,101
Bahamas	2,460	3,000 (70)			
Bahrain	2,330	649 (67)			669
Belgium	5,670	69,965 (69)	30,970 (74)	1,215 (72)	148,628
Canada	6,190	621,645 (73)	31,535 (74)	1,786 (74)	706,652
Czechos- lovakia	3,590	327,772 (74)	782 (75)	2,457 (75)	144,315

Science, Technology and Per Capita Income for Countries in the World

Countries	Per Capita GNP U.S. $ 1974	Scientists and Engineers (No.)	Establishments-Statistical Units (No.)	Employment in Manufacturing (1,000)	Enrollments in Colleges and Universities (No.)
Denmark	6,430	34,800 (70)	6,623 (74)	412.6 (74)	105,414
Finland	4,700	196,346 (73)	6,184 (74)	519.0 (75)	71,526
France	5,440	992,000 (69)		5,572 (75)	772
German Fed. Rep.	6,260	1,083,000 (73)	51,178 (75)	7,284 (75)	840,757
Greece	2,090	124,400 (69)	6,412 (73)	296.6 (73)	84,603
Hungary	2,370	336,143 (73)	1,480 (75)	1,553 (75)	103,390
Iceland	5,430	3,169 (70)			2,695
Ireland	2,320	21,884 (74)	3,211 (73)	196.7 (72)	37,897
Israel	3,460	36,000 (73)	11,192 (75)	264.6 (75)	75.338 (74)
Italy	2,820			3,590 (74)	929,300
Japan	4,070		707,804 (74)	12,024 (74)	2,155,893
Kuwait	11,770	2,151 (73)	2,650 (72)	10.3 (70)	5,800
Libya	4,640		249 (74)	10.6 (74)	11,997
Luxembourg	6,050		162 (74)	45.4 (74)	456
Netherlands	5,250	442,000 (73)		1,038 (75)	247,964 (71)
New Zealand	4,310		7,886 (73-74)	238 (72-73)	66,739
Norway	5,860	68,700 (73)	8,407 (74)	365.7 (74)	64,582
Poland	2,750	803,000 (73)	37,766 (75)	4,029 (75)	521,899 (74)
Puerto Rico	2,230		2,670 (74)	122.7 (74)	89,671 (74)

Science, Technology and Per Capita Income for Countries in the World

Countries	Per Capita GNP U.S. $ 1974	Scientists and Engineers (No.)	Establishments- Statistical Units (No.)	Employment in Manufacturing (1,000)	Enrollments in Colleges and Universities (No.)
Qatar	8,560	1,352 (74-75)			
Saudi Arabia	2,870	33,376 (74-75)			19,773
Singapore	2,240	25,234 (74)	2,405 (75)	192.1 (75)	18,195
Spain	2,490	188,000 (70)	104,139 (74)	2,158 (74)	453,389
Sweden	7,240		12,261 (74)	914.9 (74)	128,879
Switzerland	7,870		9,871 (75)		62,584
U.S.S.R.	2,600	1,939,900	39,077 (75)	29,596 (75)	4,751,000
United Arab Emirates	11,710	(73)			
United Kingdom	3,590	211,231 (68)	86,954 (72)	7,601 (74)	538,469
United States	6,670	1,614,000 (73)	299,433 (72)	18,772 (73)	10,223,729

Note: The figures in parenthesis refer to the year of the relevant data.

Appendix Sources:

Per Capita GNP: J.W., Sewell The United States and World Development Agenda 1977 (New York: Praeger Publishers, Inc., 1977), pp. 160-71.
Number of Scientists and Engineers: United Nations Statistical Yearbook 1976, Table 209, pp. 867-8, second column.
Number of Establishments-Statistical Units: ibid., Table 79, first column.
Employment in Manufacturing: ibid., Table 79, third column.
Enrollments in Colleges and Universities: ibid., Table 207, fourth column, third level of education for each country.

Notes

CHAPTER 1 NOTES

(1) On the evolution of the international system, see Falk (1975); Seyom Brown (1974); and Beres and Targ (1974).

(2) Keohane and Nye (1977); Erb and Kallab (1975); Caporaso (1978); Alker et al. (1974).

(3) Sewell (1977).

(4) American Friends Service Committee (1977).

(5) Assets owned by the largest corporations and gross annual income rival the gross national products of many states. Decisions of corporations on where to locate plants, what product lines to market or discontinue, and what technology to sell under what conditions have great importance for the economic viability of the areas in which they do business and for the future of economic growth. Barnet and Muller (1974); Apter and Goodman (1976); Said and Simons (1974).

(6) This is greater than the income of the poorest half of humanity. Two days worth of global arms spending equals the annual budget of the United Nations and all its agencies, while world research and development on armaments are six times those for energy. One-fourth of the world's scientists and engineers are connected with military R&D. Sivard (1974).

(7) Galtung, "Appraisal," in Pronk (1975).

(8) For recent reviews see Lester R. Brown (1978); Fowles (1978); Heilbroner (1976); Mesarovic and Pestel (1974).

(9) Henderson (1978); Miles (1976); Hirsch (1977). For the perspective of technological optimists, see Kahn and Martel (1976).

CHAPTER 2 NOTES

(1) UN Statistical Yearbook 1976, pp. 28 and 44, provide this information.   More precisely, the index of labor productivity in manufacturing rose from 68 in 1960 to 113 in 1975, while that of industrial production rose from 56 in 1960 to 112 in 1975.

(2) In 1960 the total exports of developed countries were 81.92 thousand million U.S. dollars. 60.12 thousand million dollars of these exports went to other developed countries. These numbers changed to 541.11 and 402.82 thousand million dollars, respectively, in 1975. The percentage of these exports has remained stationary at 74 percent. See UN Statistical Yearbook 1976, pp. 56-57.

(3) The index of per capita food production rose from 104 in 1966 to 118 in 1975 — by 34 percent, approximately 3.4 percent a year. See UN Statistical Yearbook 1976, p. 24.

(4) The term "developing countries" has itself undergone various incarnations such as backward, underdeveloped, and less developed countries (LDC). These terms are still in use, particularly LDCs.

(5) The index of per capita food production in developing countries rose from 97 in 1966 to 102 in 1975, by 6 percentage points. In other words, the per capita food production has remained stagnant in the last 10 years or so. See UN Statistical Yearbook 1976, p. 24.

(6) Productivity index in manufacturing in developing countries rose from 85 in 1960 to 101 in 1974, an increase of 20 percent in 15 years, compared to 70 percent in developed economies. See UN Statistical Yearbook 1976, p. 44.

(7) The total exports of developing economies in 1960 was 25.96 thousand million U.S. dollars. Of this, only 6.10 thousand million dollars, or 23 percent, were exported to other developing countries. The rest was exported to developed countries. The relevant quantities in 1975 were 198.08 and 48.37 thousand million U.S. dollars. The share of trade between developed and developing countries showed no change. See UN Statistical Yearbook 1976, pp. 56-7.

(8) The classification we have selected agrees with the one adopted by ODC. They, however, do not provide any rationale for this classification.   The UN uses a completely different criterion. However, UN classification is basically statistical and mechanistic. The ODC classification, on the other hand, does take into consideration, albeit implicitly, various elements in development.

(9) One can carry the game of statistical accuracy too far. What is important is to recognize the nature and limitation of the data and not to draw pointed conclusions from insufficient data.

CHAPTER 3 NOTES

(1) There is a large literature on the causes and effects of colonialism. Albert Memmi and Franz Fanon provide some of the most perceptive analyses of the colonial situation. See Franz Fanon (1966) and Albert Memmi (1967).

(2) The exaggeration of the differences is one of the important elements in exploitation.

(3) Memmi (1967), p. 117.

(4) For example, "A blonde woman, be she dull or anything else, appears superior to any brunette. A product manufactured by the colonizer is accepted with confidence. His habits, clothing, food, architecture are closely copied, even if inappropriate," Albert Memmi (1967), p. 121. Malcolm X, in his Autobiography, also provides an excellent description of such ideas.

(5) Some anthropologists and sociologists have described this phenomenon as "Westernization." The "assimilation" response has been going on for centuries.

(6) This is the genesis of many a phenomenon such as the "brain drain."

(7) This explains why the scientific research in developing countries has stayed away from analysis of traditional practices.

(8) Colonialism has encouraged the tendencies to, and characteristics of, opportunism. After all, if one can reject one's family and traditions for a little reward, what is it that one would not do for a higher reward. It has thus weakened the ethical and moral dimensions in one's actions and strengthened the selfish ones.

(9) French Canadians feel that they too are colonized. The assimilation, in their case, refers to the broader concept. "Assimilation" forms an important part of neocolonialism.

(10) We feel that this fact has been very much neglected in the understanding of development paths in the last 30 years.

(11) At present the colonial situation exists in Rhodesia and South Africa; it also exists on some small islands.

(12) One must distinguish the protracted violent struggles from the riots

and spontaneous acts of violence, by the colonized against other colonized, that did take place during the withdrawal of the colonizers from their colonies (such as Hindu-Muslim riots in India and Pakistan, Ibo-Haussa riots in Nigeria, etc.).

(13) One of the effects has been that small countries have not been able to cooperate and form regional economic units in order to reduce the disadvantages of small scale. On the contrary, they have continued, even increased, their dependence on far-off countries.

(14) Keynes's General Theory of Employment Interest and Money was published in 1936. It takes 10 to 15 years for ideas to travel from books to policy.

(15) The Harrod-Domar growth model is a simple one. It states that the rate of growth of GNP per capita can be determined by the ratio of saving propensity and capital output ratio minus the growth rate of population. This simple model has played an important part in developmental policies. Thus, India's First Five-Year Plan, which set the pace of development planning, was heavily influenced by it. See Harrod (1948) and Domar (1957).

CHAPTER 4 NOTES

(1) For example, the first five-year plan in India defined the long-term objective of doubling per capita income in 25 years by year 1975.

(2) There is a consensus about this objective among policy makers and scholars on development. The literature about this is quite vast.

(3) It is instructive to know that labor, and hence population, is treated exogenously in the Harrod-Domar model. Quite a large number of economic models follow this convention.

(4) Economists have two different meanings of distribution, how total output is distributed among the factors of production, and how the total income is distributed among people. The latter is called the size distribution of income. For welfare and development purposes, the second concept is the more relevant one.

(5) This argument is reduced to a simple form in a common home simile. The important thing is to increase the size of the pie or cake and not how to slice it. Notice how effective this simile is – slicing a pie or cake is, after all, so simple.

(6) In literature, there is recognition of the problems of distribution. However, these are all problems of the fine tuning of the system, such as removing market imperfections, taking care of special needs, and so forth.

(7) A more precise formulation would be that of constrained maximum. There is a lot of literature on the various obstacles to development. Actually, most of the literature deals with obstacles only since the authors all agree about the objectives. These obstacles may be classified in the following broad categories: 1) Resource limitations. This involves all types of resources, capital, skills, etc. See Amin (1974), Nurkse (1953), and Blaugh (1974). 2) Limitations due to increases in population. See Ohlin (1967). 3) Limitations of the economic and social structure of the society. This includes cultural issues as well as the government policies. Some of the issues are discussed in Prebisch (1964) and Little et al. (1970). We have not pursued them here since these will take us quite far. These limitations, however, do not affect the general conclusion. There is one other limitation due to technology; we pursue this in the text.

(8) More precisely, since value added is different from the value of the output, the argument holds so long as value added is a linear function of the value of output. Unless these two are inversely related, even a nonlinear relationship will not change the direction of the effect of the value of output on GNP. The exact value of the maximum, however, will be affected. We are referring here only to directional effects.

(9) In Marxian literature, the USSR is now classified as a "state capitalistic" system. However, up to the 1960s it was considered a socialist state.

(10) In standard economic textbooks, these two statements reflect the identities: $Y = C + S$, $S = I$; where Y, C, S and I stand for GNP, consumption, savings, and investment, respectively.

(11) Resource gap follows from the following. To achieve a certain level of rate of growth of GNP, a certain level of investment is required, say $I*$. If the actual savings (S) are less than the desired level of investment, there is a gap, i.e., $I* - S$. This is the resource gap. See Chenery (1966).

(12) The liberal position in development theories has been that some poor countries cannot save enough and thus fall in the vicious cycle of poverty. The argument runs thus: low income − low savings − low investment − low production − low income − low savings. This thesis is best articulated by Nurkse (1953). The onlyway out is either a strong regimentation or foreign aid.

(13) It is not a simple problem. The Prebisch thesis deals basically with this problem. The thesis is that poor countries export their raw materials in a competitive market while they import the industrial goods in a monopolistic market. In the process they lose quite a fraction of their purchasing power. See Prebisch (1964).

(14) A recent UNCTAD study has suggested that brain drain represents a large transfer of resources from the developing to developed countries.

(15) Galbraith elaborates on this thesis for United States corporations. See Galbraith (1965).

(16) This process is going on even in the highly monetized societies. In the developed countries money values are now being assigned to time. For an interesting analysis in this regard, see Linder (1970).

(17) Theoretically, there is always an implicit resource or real price for any nonmonetized exchange. However, it is not at all clear what this price is. The introduction of money relations changes the social relations and the character of exchange so fundamentally that the situations prior to and after monetization are not comparable. There is no easy way to compare this real price with the "money price" that follows from the introduction of exchange relations.

(18) Here a distinction is made between the price of one commodity which is determined by supply and demand and the matrix of all the prices. Income distribution determines the matrix of all prices.

(19) On closer examination, one can find in this ideology similarities to the arguments made by colonizers who also justified their privileges on the basis of their work effort and the poverty of the colonized on their laziness.

(20) There has been a controversy between classical and two-gap theorists. One of the fundamental elements of two-gap theory is that imports are an essential input in production; i.e., the production function is $Y = f(K,L,M)$ where Y, K, L, M are output, capital, labor, and imports. See Diwan (1968).

(21) For an excellent exposition of this variant, see Felix (1966).

(22) The concept of production capacity is a tricky one. There is quite an amount of literature on what constitutes "capacity."

(23) The trade theory argues that all trade leads to gains for every trading partner. However, it assumes that trading partners are equal. In case the trading partners are unequal, the gains do not follow. See Diwan (1973a).

(24) The Brazilian Model (BM) needs to be distinguished from the Export-Led-Growth (ELG) model. BM involves an emphasis on export of manufactured goods while ELG involves emphasis on those from the primary sector. However, the differences are much larger than this. ELG implies positive backward linkages via increases in saving, investment, employment, and hence incomes. At least in theory ELG ensures development, even if in the very long run. In the BM there are no linkages; there are thus no increases in savings, investments, and employment. This is so because the very purpose of BM is to make use of unused capacity generated through government subsidies in the first

place. ELG generates a process that reduces dualism in the society, at least in theory. BM is the end of the process that leads to dualism. BM hardens the dualism process into a nonfunctional dual society. An important component of BM is political repression which follows from the need of subsidies to the industrial sector, and maintaining a nonfunctional dual society.

(25) There has been a continuous, and unresolvable, controversy between the liberals and conservatives about the effectiveness of "foreign aid" and "private foreign capital." Liberals have argued for foreign aid and conservatives have found advantages only in "private foreign capital."

(26) Many scholars argue that agricultural production under the "green revolution" resembles industrial production. See Wharton (1969) and Frankel (1971)

CHAPTER 5 NOTES

(1) The production of rice in East Asian societies provides a relevant example.

(2) There are other considerations such as the emphasis on what Illich calls "conviviality" instead of productivity, human interaction, etc. See Illich (1973).

(3) See Ellul (1964). Ellul uses the term "techniques" instead of "technology."

(4) It must be added that this sentence is not exclusive. It does not mean that science and technology do not play an equally important part in other development strategies.

(5) Specifically, if $Q = Af(K,L)$ where Q, K, L represent output, capital, and labor. A is an indicator of technical change using this function. Solow estimated that technical change explained 90 percent of the growth in output in the United States between 1909 and 1949. See Solow (1956).

(6) This concept is akin to "technical efficiency." However, "technical efficiency" has to be distinguished from economic efficiency which involves both prices of inputs and the quantities of the product. "In practice, technical changes in products often embody all three types of change simultaneously; that is, the new products offer greater need fulfillment in relation to resource cost, change in the balance of characteristics and embody greater quantities of resources." Frances Stewart (1977), p. 17. For the "technical" concept, see Eckaus (1977).

(7) Embodied technical change results from new capital. Disembodied technical change is represented by shifts in the production function.

There is a serious question if disembodied technical change can be defined. Griliches and Jorgensen (1967) were able to explain all productivity by the embodiment hypothesis.

(8) See Ellul (1964).

(9) Commoner considers it the fourth and perhaps the most important law of ecology. It is related to the first two laws: "everything is connected to everything else," and "everything must go somewhere else." See Commoner (1971).

(10) This concept has been used by Ellul (1964), Goulet (1977), Illich (1973) and Stewart (1977b) to name a few. "Thus, a much broader definition (of technology) is adopted, extending to all the skills, knowledge and procedures of making, using and doing useful things. Technology thus includes methods used in nonmarketed activities as well as marketed ones. It includes the nature and specification of what is produced — the product design — as well as how it is produced. It encompasses managerial and marketing techniques. . . ." Frances Stewart (1977b), p. 1.

(11) When high yielding seed varieties and fertilizers were used in Indian agriculture, there was a major increase in wage-labor so that there were statements to the effect that "Green revolution leads to red revolution." See Wharton (1969) and Frankel (1971).

(12) There is ample evidence on this proposition — one can obtain it from any country from the type and nature of goods produced and sold. Political and bureaucratic needs translate into prestigious goods. Business needs are reflected in profitable goods.

(13) In the 1950s there was a lively discussion in India about the emergence and growth of what was called the "U sector" involving such goods as luxury housing.

(14) This not a solitary example. India built a whole complex of buildings to host a UNESCO meeting in the 1950s. Mexico spent millions for an olympic meet. Montreal is saddled with the cost of an olympic stadium and the Canadians are still paying for it.

(15) There are estimates that as much as 95 percent, and even more, of total R&D expenditures are in developed countries.

(16) There is ample literature and evidence on the propositions that government contractors have pushed particular (even if not useful) costly technologies to the government. Similarly, the multinational corporations are known to have exported wrong, even harmful, technologies to other countries. Nestle's baby food is an interesting example. There are studies that show that United States multinationals have exported hazardous goods and technologies after these have been banned in the United States.

(17) In the past few years a number of United States multinationals have admitted bribing the buyers in other countries. The corporation spokesmen have even argued that bribing is normal practice, particularly in developing countries. However, the bribing has not been limited to developing countries only. It has been reported that Lockheed may have bribed decision makers in the Netherlands and Japan.

(18) The convention has been to define efficiency in terms of the first law. This concept of efficiency is more suited to the technique concept of technology. The process concept of technology bears greater relationship to the efficiency based on the second law.

(19) Productivity has generally been defined as output-labor ratio. However, productivity is also defined as a ratio between output and a weighted sum of various factors. The inclusion of factors and the assigning of weights open up many possibilities.

(20) To name only a few, there is demand price, supply price, monopoly price, price based on costs, price that will replace the machine, etc. All these prices are different. It is only in very well defined markets, such as perfect competition, that these prices are uniquely related.

(21) See ILO (1977), pp. 51-6.

(22) ILO (1977), pp. 162-7. Also see Management Analysis Center (1976).

(23) See UNCTAD (1972) and (1974).

CHAPTER 6 NOTES

(1) There are many scholars who explain the growth and development of these countries on the basis of exploitation of Third World countries via colonialism, imperialism and neocolonialism. See Andre Frank (1970) and Samir Amin (1974). Our arguments in terms of CDS are not inconsistent with these explanations. However, our arguments are based on much different factors and influences.

(2) Kenneth E. Boulding, "The Economics of the Coming Spaceship Earth," in Jarrett (1966), pp. 3-14.

(3) For recent overviews of the United States energy situation, see U.S. Executive, The Domestic Council (1976), pp. 7-11; Norman (1978); Brown, N. (1978).

(4) Grossman and Danekar (1977); Henderson (1978); Hannon (1976); Lovins (1977); Daly (1977); Hayes (1977), pp. 140-51.

(5) Ford Foundation (1974).

(6) The changes include switching from planes and cars to intercity trains, throwaway to refillable beverage containers, a new highway construction to health insurance and personal consumption, and cars to buses and bicycles. Hannon also states that in 1974 a drop in energy demand at the rate of 930,000 per quad correlated with a gain in new jobs. Hannon (1976), pp. 8-9.

(7) Harman (1978).

(8) Lovins (1977); Hayes (1977); Brown, N. (1978), pp. 178-82.

(9) Schumacher (1973); Henderson (1978); Stein (1974); Fritsch (1976).

(10) Lovins (1977); Ophuls (1977).

(11) U.S. National Center for Appropriate Technology (1976). The percentage of poor families subject to shutoffs by utilities has been increasing.

(12) For an excellent analysis of the costs of growth, see Mishan (1970).

(13) For an analysis of alienation, see Weisskopf (1973).

(14) Linder (1970) provides an illuminating analysis of how scarcity of time is changing culture.

(15) Packard (1972) persuasively demonstrates how mobility is breaking up communities and leading to more crime and decay.

(16) Heilbroner (1974). Heilbroner feels that this civilization malaise may lead to the decline of the business of civilization.

(17) He gives the equation $E = (E/Y)(Y/N)N$ where E, Y, and N refer to environment degradation, GNP, and population, respectively. E/Y is the impact of technology. Y/N is affluence and N population. He provides evidence on the statement that changes in E are related to changes in E/Y. See Commoner (1971).

(18) The nitrites in bacon and heating of hamburgers at high temperatures are considered carcinogenic.

(19) Only a few years ago one heard that Lake Erie was dying by "eutrophication." To save it involved a costly and long process.

(20) The production of manufactured goods in developing market economies over the period 1960 to 1975 has grown by 260 percent. These figures are from the UN Statistical Yearbook 1976.

(21) The consumption of energy over the period 1967 to 1975 has increased by 169 percent in Africa, 154 percent in Asia excluding

Middle East, and 171 percent in South America. Over the period 1960 to 1975, electricity production has increased by 460 percent in developing market countries. The production of gas and petroleum has increased by 313 percent, and coal by 178 percent. These data are taken from UN Statistical Yearbook 1966.

(22) The index of food per capita in developing market economies was 102 in 1975; 1961-65 = 100 in 1961 to 1965. UN Statistical Yearbook 1976.

(23) ILO estimates that there are 1,210 million seriously poor and 706 million destitute in 1975 in the developing market economies. They form 67 and 39 percent of the population, respectively. Their number has increased by 119 and 43 million respectively in the period 1963-1972. See ILO (1977), pp. 22-3.

(24) ILO estimates that 40 percent of the working force was either unemployed or underemployed in developing economies in 1975. ILO, ibid.

(25) These conclusions follow logically from the facts in A, B, and C. Diwan (1978c) provides details on this for India.

(26) This conclusion is of major importance; it raises fundamental questions about development.

(27) The basic theory of a dual economy is given in Jorgensen (1961) and Fei and Ranis (1964). Since then there have been a number of refinements in the argument. The basic structure of the theory, however, has not changed.

(28) The fifties and early sixties were the heady days of conventional development strategies when economists believed they could recreate the affluence of the West in the poor countries in a matter of decades. It was thus axiomatic to assume that all the society and people were waiting to move to the modern sector, which naturally was "heaven."

(29) The distortions have been described in terms of market imperfections caused by lack of supply of capital and skill formation institutions and by the distorted factor prices created by tax and subsidy policies of the government.

(30) There is ample literature here by Marxian economists starting with Karl Marx himself. Representative members of the school are Samir Amin (1974) and Andre Frank (1970).

(31) Sethi (1975) has attempted an answer to this question, albeit implicitly. His answer is that these objectives are in conflict and that the dynamics of the dual society will eventually lead to the very elimination of dualism via revolution. This is the real implications of his

phrase, "nonfunctional dual society."

(32) For example, one can appreciate the continuous eroding of the goals of political power expressed by the people in the overthrow of the Indira Gandhi government in the 1977 elections. The Janata Party and government has steadily lowered its commitment to Gandhian principles. Some of these ideas are explained in Diwan and Bhargava (1978b).

(33) See Chapter 3.

(34) Stewart (1977a) defines these as high income products. One has an uneasy feeling about such a term since it implies that high incomes naturally involve these products. There is a need for a better term.

(35) The distinction between "hard" and "soft" technology paths has become crucial these days. See Lovins (1977).

(36) There are many elements in this hypothesis that are similar to the arguments of CDS. Thus the emphasis on GNP growth may coincide with, and be the result of, capitalist modes of production. As we have defined the centralized modes of production, these are not different from capitalistic modes. However, the difference is that some production modes may be capitalistic in terms of "private ownership" and not "centralized."

(37) This hypothesis is best articulated in the writings of Professor David Felix, Washington University, St. Louis. One of the authors, Romesh Diwan, is thankful to him for previous discussions on this topic.

CHAPTER 7 NOTES

(1) Sen (1976) suggests an interesting name for this transition: "pure exchange systems transition" – PEST for short. In the system prior to this transition, income and production were integrated. The effect of this transition is to separate these two. There are thus two problems: people who do not and cannot produce, and people who do not or cannot have incomes. "The system of labor and income prevalent in the PEST phase seems not merely to increase vulnerability to disasters, but also constrains dramatically the country's ability to make use of available resources. The inhumanity of the system competes with its equally distinguished inefficiency."

(2) For example, in the case of India, Dr. Minhas writes, "In spite of a number of references in the five year plans to employment problems, the creation of employment opportunities was seen more or less as an adjunct to or a by product of the development strategy." Cited in Diwan (1977b).

(3) For analysis of the two-gap theory, see Chenery and Strout (1966) and Diwan (1968b).

(4) The pressure to secure aid still persists. All books on development automatically include a chapter on "aid" (e.g., the Overseas Development Council's annual volume on United States and World Development has a regular section on aid).

CHAPTER 8 NOTES

(1) Diwan (1977b) argues that one of the important elements in the overthrow of the Indira Gandhi government in India in 1977 was the impression by the people that the existing development strategy was harmful for them and needed a major change. The opposition parties did talk of an alternative development strategy. A variant of ADS has been followed in China. See Gurley (1976).

(2) The limits to production of materials come from political, energy, and environment considerations. In the last analysis, ecosphere provides the ultimate limit.

(3) This point is now being made in all sorts of circles. It is not a political plot by the rich countries against the poor countries. It is an expression of the recognition that the ecosphere cannot withstand such high levels of industrial production.

(4) Some scholars argue that this relationship is of the form of an invested parabola: there is a maximum level of welfare and beyond that level increases in GNP reduce, instead of increase, welfare. Thus, after the maximum point, a meaningful objective may be to reduce GNP.

(5) Samuelson (1976).

(6) Sen (1976). Writing e, Y, and G for life expectancy, GNP per capita, and gini coefficient, respecitvely, the index is e Y (1-G).

(7) John W. Sewell et al. (1977).

(8) In the last part of 1978, Iranian people started a reaction to the Iranian government's repressive policies. Shah was ousted and an Islamic Republic has been formed. It has created worries in the corridors of the United States Department of State.

(9) See Seers (1969) and Huq (1976).

(10) For details on these needs see Maslow (1962). He distinguishes between basic and meta needs. The interesting thing is that his hierarchy of basic needs contains a large number of nonmaterial needs. By and large, all these nonmaterial needs depend essentially on the existence of a community of human beings and not a community of interests. Emphasis on material production cuts at the very root of a community.

(11) ILO (1977).

(12) Janata Party Economic Policy Statement. Janata is the ruling party in India.

(13) Research Policy Program (1978).

(14) Schumacher (1973).

(15) See Rath and Dandekar (1971).

(16) See Diwan (1971). Calorie defined poverty in India implies that 40 to 50 percent of the population are poor. However, as Diwan suggests, a small dose of medicine and cloth raises the level of poverty to the level of two-thirds of the population in India.

(17) See ILO (1977), p. 22. ILO distinguishes between the seriously poor and destitute. The destitute form a hard core of the seriously poor and are two-fifths of the population. Their consentration is highest in Asia and Africa — more than two-thirds of the population.

(18) ILO (1977), p. 18.

(19) The overall estimates can be judged from the following facts. The United States population is 200 million. The total world population of the poor is 1,200 million, just six times the population of the United States. The per capita resource consumption in the United States is more than six times the basic need.

(20) This is an effect of colonialism and a repressive system in which there is little hope and capacity to develop one's own solutions.

(21) See ILO (1977).

(22) Diwan (1977a) distinguishes between "self-defined" and "stranger-defined" work. The work most relevant to the development of a human being is the "self-defined" work. Self-defined work is the work defined by the person who works, or by persons who share in this person's job, grief, and suffering, such as family members. A very large part of work in an exchange economy is stranger-defined.

(23) Boulding (1949-50) argues that welfare should be defined as a function of stocks instead of income. Thus, a person is better off the more stock he/she holds, and not the higher the income. The concept of stock used is that of an economist's in terms of physical quantities and not in terms of money quantities.

(24) Chenery et al. (1975). See also the Economic Policy Statement of the Janata Party in India; and Diwan and Bhargava (1978b).

(25) Goulet (1978).

(26) Classical economists have distinguished between values-in-use and values-in-exchange. Basic needs goods reflect values-in-use. GNP, on the other hand, emphasizes values-in-exchange. Neoclassical and Neo-Marxian economists have, unfortunately, ignored this vital distinction, thereby losing a powerful analytical tool.

(27) A number of protagonists of CDS are misinformed about the growth potential of ADS. For example, see R. Eckaus (1977).

CHAPTER 9 NOTES

(1) See Chenery et al. (1974).

(2) See Hayter (1971) for details.

(3) Sinha et al. (1978), pp. 6.15-6.16.

(4) See ILO (1977).

(5) This is perhaps the most serious problem with intergovernmental, and bilateral, institutions. These bodies do talk about the "relevant" issues. They can also spell out, if allowed, the implications of the solutions. Yet they never bring forth these implications. For political and bureaucratic reasons, they stay away from these solutions and thus search for nonsolution solutions. In the process, these institutions do a disservice, even harm, insofar as they encourage complacency. In this respect nongovernment organizations can play a much more important role.

(6) This manifesto has been, understandably, hurriedly written. It provides the basis, and an outline, of a completely new strategy. Yet it has enough loose ends, repetitions, and inconsistencies to be miscon-strued, if one is bent upon giving it an uncharitable interpretation.

(7) We have grouped and regrouped the various salient features to make these as consistent as possible. Our listing is not necessarily compre-hensive. It hopefully captures the essentials in the manifesto.

(8) Janata Economic Policy Statement. Quoted in Diwan and Bhargava (1978b).

(9) The economic and social policy in China has changed fundamentally since the death of Mao. Charles Bettelheim (1978) has argued, persuasively, that the party line after Mao is nonrevolutionary and positively bourgeoisie. He points out some of the mistakes Marxist's make by not recognizing this major shift in the policy in China.

(10) See Diwan (1977b).

(11) Gurley (1976), p. 13. The book develops the thesis in detail and provides extensive evidence on the Maoist strategy of economic development.

(12) We are aware that these comparisions are rather rough. The historical conditions in the two countries were rather different. Our interest in comparing them is simply to highlight the point that CDS and ADS are relevant to socialist economies insofar as the objectives of CDS and ADS are so different.

(13) Nyerere (1974) p. 26.

(14) Ibid. p. 37.

(15) Goulet (1978), p. 1. This book contains the details of the development process in Guinea-Bissau.

CHAPTER 10 NOTES

(1) See Reddy (1975). There are certain technologies which help only the poor and are of no use to the rich, such as spray of mud walls.

(2) For a cogent formulation of this argument, see ibid.

(3) On attempts to characterize appropriate technology, see Schumacher (1973), Godfrey Boyle et al. (1976), Nicolas Jequier, ed. (1976), Robin Clarke (1977), and Canadian Hunger Foundation (1976).

(4) See Barry Stein (1974), Hazel Henderson (1978).

(5) Jequier, ed. (1976), pp. 15-20.

(6) Ibid., p. 7.

(7) See Eckaus (1977).

(8) It should not be surprising since the N.A.S. views the idea of development from the point of view of us, the developed and them, the developing.

(9) These are discussed in Chapter 2.

(10) The definition was worked out at a lunch by Denneth Dohlberg, Romesh Diwan, Ted Owens, and Michaele Walsh at an NSF-supported workshop in Troy, New York, July 10 to 12, 1978, organized by Romesh Diwan and Dennis Livingston.

(11) This is particularly true of the definition adopted by the UN. It is pure gobbledygook, full of bureaucratic jargon.

(12) Congressional Record, vol. 124, No. 24, February 24, 1978.

(13) Vance Packard argues that mobility is particularly disruptive to a community. See Packard, 1972.

(14) Lovins (1977). His soft energy paths involve community defined technologies in the harnessing of energy from the various forms of solar energy. His cost estimates clearly show that hard technologies based on oil, coal, and nuclear power are inefficient.

(15) Brown and Howe (1978), pp. 651-2.

(16) See reports on the Task Force on Integrated Rural Development, Government of India. Professor Minhas was the chairman of this task force.

(17) The argument advanced in favor of these technologies is that they save food from waste. This argument, particualrly in countries with sizeable poor populations, is false. In such countries, no food available to the poor for consumption is _ever_ wasted. In fact, waste occurs because of the social and eocnomic system – such as private profit – by which the poor are denied access to these foods.

(18) This point was suggested to one of the authors, Romesh Diwan, by Devinder Kumar.

(19) We are aware of difficulties of measuring long-run social costs. It calls for intellectual and scientific effort in this direction.

(20) The result will look somewhat like the following. The objective numbers on the right, however, will be different. The table below is purely an example.

Priorities for Technology Choice on the Basis of ADS Objectives

Priority	Ability to Achieve ADS Objectives
1st	All five objectives
2nd	Objectives 1, 2, and 3
3rd	Objectives 1 and 2
4th	Objectives 1 and 3
5th	Objectives 2 and 3
6th	Objective 3
7th	Objective 2
8th	Objective 5
9th	Objective 1
10th	Objective 4

(21) The final table will look like the following. The numbers on the right will be different. This is only an example.

Priorities for Technology Choices on the Basis of
Technological Characteristics

Priorities	Technology Characteristics
1st	all seven characteristics
2nd	characteristics 1, 2, 3, and 4
3rd	" 1, 2, 3, and 5
4th	" 4
5th	" 5
6th	" 2
7th	" 1
8th	" 3
9th	" 6
10th	" 7

CHAPTER 11 NOTES

(1) Winner (1977); Dickson (1977); and Bereano (1976).

(2) In the 1930s and 1940s, these included Ralph Borsodi and Helen and Scott Nearing (still demonstrating the benefits of simple living today).

(3) Morrison (1978).

(4) These include the Institute for Local Self-Reliance, Institute for Policy Studies, and Community Technology, Inc. (all in Washington, D.C.); the Farallones Institute (Northern California); and the New Alchemy Institute (Massachusetts). There is an Institute for Appropriate Technology at the University of California at Davis.

(5) For example, The CoEvolution Quarterly (Box 428, Sausalito, CA 94965); Rain (2270 N.W. Irving, Portland, OR 97210); Alternative Sources of Energy (Rt 1, Box 90A, Milaca, MN 56353); Appropriate Technology (IT Publications, 9 King Rd., London); TRANET (Transnational Network for Alternative/Appropriate Technologies, Box 567, Rangeley, ME 04970); New Roots (University of Massachusetts, Amherst, MA 01003).

(6) An intuitive list, based on media exposure and references in the literature, might include Karl Hess, Murray Bookchin, Wendell Berry, John and Nancy Todd, Hazel Henderson, Stuart Brand, Amory Lovins, Barry Commoner and, before his death, E.F. Schumacher.

(7) Manufacturers of solar energy have formed one such association.

(8) For example, educational programs are run by Domestic Technology, Inc. (Colorado), the Shelter Institute (Maine), Total Environmental Action (New Hampshire), and the School of Living (Pennsylvania).

(9) In the United States, these include India Development Service, Volunteers in Technical Assistance (VITA), and Volunteers in Asia. In other developed countries, there is the Intermediate Technology Development Group (Britain), Brace Research Institute (Canada), TOOL (Netherlands), and GRET (France). Together with VITA, these last four organizations have formed an informal International Network for Appropriate Technology.

(10) In the United States there are Offices of Appropriate Technology in Lane county (Oregon) and Santa Clara County (California), and in the state government of California. The Federal government has estalbished a National Center for Appropriate Technology oriented to low-income communities, while AT research and development are funded by the Natonal Science Foundation, Department of Energy, and other agencies.

(11) For example, the Toward Tomorrow Fair, held annually at the University of Massachusetts, Amherst.

(12) These include the Meso-American Center for Studies on Appropriate Technology (Guatemala), Northern Technical College (Zambia), Appropriate Technology Development Organization (Pakistan), Appropriate Technology Resource Service (Barbados), University of Science and Technology (Ghana), and Office of Village Development (Papua, New Guinea). Lists of such groups may be found in Singer (1977), and the attachments to the U.S. Agency for International Development (1977).

(13) Canadian Hunger Foundation (1976), Darrow and Pam (1976), Editors of Rain (1977), Congdon (1977), Jequier (1976), Bulfin and Greenwell (1977), Boyle and Harper (1976).

(14) In general, see Stewart (1977a).

(15) Reddy (1975).

(16) International Labor Organization (1978).

(17) Bulfin and Greenwell (1977), p. 6.

(18) Bernstein (1978).

(19) U.S. Congress, Office of Technology Assessment (1977), pp. VII-25. This study notes that long-term implications of technological development for employment cannot be reliably determined with "contemporary economic methods." A study of the employment implications of an aggressive solar energy program in California revealed that 377,000 jobs a year could be created in the 1980s, thereby cutting in half California's

present unemployment total(Branfman and LaMar (1978)). Another study by James Benson of the Council of Economic Priorities estimates that in Nassau and Suffolk counties, New York, an energy conservation program plus solar hot water systems would provide 270 percent more employment and 206 percent more energy over a 30-year period than the amount of money spent on constructing and maintaining nuclear power stations.

(20) Wakefield and Stafford (1977).

(21) "Community economic development is narrowly defined as those political and economic activities undertaken by institutions, created and controlled by community residents, that lead to a sustained growth in the net creation of quality employment and per capita income as well as promote equity of ownership and personal income." Yaksick (1978), p. 1.

(22) Morris (1977), Morris and Hess (1975).

(23) Smith, Miranda, (1978).

(24) Stanley (1978).

(25) Lovins (1977), U.S. National Academy of Sciences (1976a and 1978).

(26) Canadian Hunger Foundation (1976).

(27) Commoner et al. (1975); see also Merrill (1976) and Berry (1977).

(28) Goldschmidt (1978). The quality of life in the family farm town was high because local institutions received greater support and money was spent in the community.

(29) Cited in Balis (1976).

(30) Robertson (1978). This community, however, has encountered financial and leadership problems. The Farm, a large commune in Tennessee, is already a working model of a self-sufficient community and, in fact, operates its own technical assistance agency, called Plenty.

(31) Lappe and Collins (1979).

CHAPTER 12 NOTES

(1) For an excellent analysis of the logic of industrial modes of production see Illich (1973). He explains how industrial modes change and redefine needs, and how these modes create and accept only consistent institutions.

(2) Research effort on these issues will constitute what Illich calls "counterfoil" research.

(3) There are cases when some people may not be able to identify their needs (such as drunks). These are, however, exceptions. There are other cases, which are more serious, when certain groups may define some of the privileges as their basic needs.

(4) If a person fails to obtain "access" from these two institutions, and has no money of his/her own, the only means left is stealing, even "mugging." In the United States it is not uncommon to find "mugging" used to satisfy basic needs. However, penalities are very high and it is not a sensible solution by any means.

(5) In history, this relationship has been formalized into serfdom and slavery. By providing "access," the landlords managed to even own the human being. The exploitation continues today.

(6) Some scholars distinguish between feudalistic and capitalistic modes of production. They define landlordism as feudalistic only. For many questions, and important ones, this distinction is useful. For our purpose, this distinction is not that important. What is important is whether everyone has an access to means to satisfy needs; in this case land is a means.

(7) Our interest here is in community control and not how the land is given to the community. The community can purchase the land, appropriate it, or obtain it from the state which has nationalized it or obtained it by other means. There is no doubt that the way land falls into the hands of the community will have some effect. In the long run, this effect is negligible.

(8) Some of these issues, on distribution, are developed in Diwan and Gidwani (1978d).

(9) In production there are economies of scale so the number of persons increases at a smaller rate. In bureaucracy, on the other hand, the number of bureaucrats may increase at a higher rate.

(10) For want of better terms, we consider these two processes in their broadest meanings so that production includes distribution as well as the production of services – virtually all economic tasks. Similarly, administration is intended to contain all political and bureaucratic tasks. Hopefully, production and administration covers all the major economic and political activities.

(11) There have been constant references to the effect that bureaucracy is the fastest growing industry.

(12) We now have such offices as deputy assistant to the undersecretary,

and assistant to the deputy of the assistant secretary. It seems there are all sorts of permutations and combinations of the words, assistant, deputy, associated, and under.

(13) These bastions of free enterprise never get tired of rallying against bureaucracy.

(14) The funny thing is that the International Monetary Fund and World Bank in their various operations continuously harp on free market philosophy, complaining against bureaucracy. Yet they are able to promote bureaucracy within their own organizations. One presumes that if one has power, one can justify all inconsistencies.

(15) The reward structure, particularly inordinately high rewards to top layers, has been rationalized on various grounds such as higher intelligence, higher efficiency, higher capacity, higher human capital, higher productivity, etc. None of these can stand close analysis. The only rational explanation is higher political and economic power, and yet this explanation is not offered.

(16) The effects of hierarchy have strong similarities with the colonial situation. See Chapter 3.

(17) The debilitating and dehumanizing impact of bureaucracy is best portrayed in Kafka's famous novel, The Trial.

(18) Actually, this has made the hierarchical structure virtually obsolete. Although the president of a corporation may have all the power, he/she cannot fully understand each and every process going on. The attention span is so low that one cannot understand, and comprehend, even half complex issues. These issues have to be boiled down to simplicities.

(19) Committees make better decisions since members of the committees can contribute a lot of diverse information to the issues. The committee has a built-in structure for conflict resolution, the democratic process. Though committees have been rallied against because of obsolete ideas, they are even more efficient mechanisms. Interestingly, no important decisions have ever been made without committees. In law, the most important cases are decided by a committee of judges.

(20) Networks may also be considered as an extension of the committee system.

(21) The difference between the committee and hierarchical system is similar to the difference between democracy and dictatorship.

(22) In the old days these decisions used to be about life and death and could be made by king/queen. Even now the same idea is being followed. Thus, the president of a country, say the United States, has the power

to decide about the use of nuclear arms thereby killing millions of people. Of course the king is always right. No wonder the president is on the top of the ladder of hierarchy.

(23) It will also have the salutary effect of keeping tendencies towards centralization in abeyance. Instead, it will lead toward decentralization.

(24) "Supposing I have come by a fair amount of wealth by way of legacy, or by means of trade and industry, I must know that all that wealth does not belong to me, what belongs to me is the right to an honorable livelihood, no better than enjoyed by millions of others. The rest of my wealth belongs to the community and must be used for the welfare of the community. . . ." M.K. Gandhi (1957) Vol. I, p. 157.

(25) There are various debates about what is meant by democracy. We do not wish to join in this debate. By democracy we mean the following: freedom of information, speech, and assembly; equality before the law; the right to periodically elect, and reject, representatives who are expected to make decisions on people's behalf; and the right to participate in fulfillment of basic needs.

(26) Many times the statement is made that democracy is a luxury for the poor. All they need is economic improvements; they do not care for democracy. This argument is based on an inverted and twisted logic. The experience of elections in India in 1977 has proven, hopefully once for all, the fallacy of the assumption that people do not care. People do care. Furthermore, dictatorships are not more efficient. They can give the appearance of efficiency. After the murder of Allende, president of Chile, and the takeover of the Chilean government by military junta, the prices in the first year rose 1,500 percent while the wages were frozen. Economic conditions have not improved since.   Lack of democratic practice has been one of the most fundamental weaknesses in the otherwise socialist states. Some scholars argue that this factor has been responsible for the rise to power of the bourgeoisie leadership in China since the death of Mao.

(27) Panchayati raj means rule by five elders of the village, generally elected on the basis of their care for the community and the respect people have for them.

CHAPTER 14 NOTES

(1) Robertson (1978). For a comprehensive analysis of differing connotations of postindustrial, see Marien (1977).

(2) For expressions of the conserver society stressing its decentralist, steady-state characteristics, see Goodman (1977); Henderson (1978); Satin (1978); Valaskakis (1978); Pirages (1977); Ophuls (1977); Daly (1977); Lovins (1977); Nash (1978).

3) We recognize that there are severe gaps in standard of living within developed countries equivalent to those between rich and poor nations. Just as developing countries must increase their share of and access to world resources, the same is true of groups and sections in developed countries who have traditionally lived at poverty levels. We do not underestimate the political difficulty facing developed countries that attempt to expand the well-being of such groups at a time when the overall economic pie grows ever more slowly.

(4) Livingston (1977).

(5) In the United States, the Department of Energy runs a small grants program in Appropriate Technology that attempts to address these needs, as does the efforts of the National Center for Appropriate Technology. The National Science Foundation has held a series of regional forums throughout the country to help it formulate an AT program, although the applied technology aspect of such a program is in conflict with NSF's traditional commitment to pure science. As with AID, NSF's initiatives here have come as a result of congressional legislative pressure.

(6) That is, solar and other sources of renewable energy are not competing on fair grounds in the marketplace with subsidized sources. Subsidies include oil production incentives, oil depletion allowance, government funding of nuclear R&D, and tax loopholes for utilities. Subsidies encourage energy growth in general, as well as from inappropriate sources. Battelle Memorial Institute (1978) and Brannon (1977).

(7) Lovins (1977). International developments in soft energy path analysis and policies are reported in Soft Energy Notes (International Project for Soft Energy Paths, 124 Spear St., San Francisco, CA 94105).

(8) Hayes (1978).

(9) Studies by the Council on Economic Priorities and Chase Econometrics Associates are cited in the New York Times, November 19, 1978, Section III, p. 1,6.

(10) These policies are only a representative sample. More drastic moves are conceivable. Members of the Secretariat for Future Studies, a Swedish government agency, have proposed maximum consumption levels on meat, a ceiling on oil consumption, the matching of buildings more evenly with family size, increasing the durability of consumer goods, and abolishing the private ownership of automobiles. Backstrand and Ingelstam (1977).

(11) Stavrianos (1976); Stokes (1978); Morris and Hess (1975).

(12) Burns (1975). This author also claims that assets of households

totaled more than a trillion dollars, producing an annual return almost equal to the net profits of every corporation in the United States.

(13) Morehouse (1977). Morehouse also suggests that the United States reexamine corporate tax policies which subsidize direct investment abroad, and that in readjusting the types of technology that are encouraged or discouraged for export, consumer technologies for the urban minority receive low priority while technologies for processing raw materials receive higher attention. Shields (1977-78) argues for the slowing down of transfers for a few years, though not for permanent barriers to foreign licensing.

(14) Management Analysis Center (1976).

(15) Development Forum, March 1978. The UN draft refers to the need for MNCs to provide information to host governments on prohibitions or warnings imposed in other countries where the MNC operates, on health and safety grounds, concerning products the MNC markets in the host country, and information on the effects of products or production processes.

(16) Development Forum, October 1978. A major controversy is whether or not this code should be binding or, as developed countries prefer, take the form of suggestive guidelines. Many other aspects of the code, including machinery for implementation, means of settling disputes under it, and the meaning of technology transfer itself are in dispute.

(17) Section 102 (b) of Foreign Assistance Act of 1961, as amended.

(18) The chairman of OPIC is the administrator of AID. OPIC projects focus on agribusiness, minerals, and energy production; Congress has mandated OPIC to become more involved in low-income developing countries, to increase the proportion of small business investment projects insured to 30 percent of the total, and to give loans to projects significantly involving small businesses or cooperatives.

(19) Weisman (1970).

CHAPTER 15 NOTES

(1) If one remembers that Latin American governments were highly influenced by the United States, one can easily estimate the influence of the United States in the UN. In view of this influence, the United States could keep People's Republic of China out for a long time.

(2) Some scholars suggest that the future of these institutions depends heavily on the policies of countries like Saudi Arabia. They speculate that the nature and composition of both World Bank and IMF will change drastically in the next 10 years, as the economic power shifts to these

untries and they learn to use it for their objectives.

) There is now ample evidence on the proposition that the international economic system has resulted in the transfer of resources from
the developing to developed countries. See Prebisch (1964) and Diwan
(1973a). Prebisch explains this on the basis of a free market in raw
materials and monopolies in manufactured goods. Diwan provides an
explanation in terms of unequal trade partners.

(4) The emphasis in UNCTAD I in Geneva was to secure "aid" from the
developed countries. It was suggested that developed countries should
provide 1 percent of their national income in forms of aid to developing
countries. This target was never achieved and now it has been reduced
to .7 percent. There is no reason to expect that it will be achieved.

(5) See Chapter 2.

(6) We have explained this in terms of dual society. See Chapter 6.

(7) It was such an important act that there was talk, in many developed
countries, of a military action against some of these countries. In the
case of the United States, their involvement in Vietnam had already
made the people weary of war. Some scholars argue that the
Vietnamese war was partly responsible for the success of the increase in
the price of petroleum.

(8) The literature on the impact of the transfer of resources to OPEC is
mushrooming. This has become an important element in any study in
international relations and issues.

(9) This is a UN General Assembly resolution and is available from the
UN. A useful source is the UN Monthly Chronicle, April 1974, p. 66.

(10) Ibid. pp. 67-9.

(11) "To give to the developing countries access to the achievements of
modern science and technology, to promote the transfer of technology
and the creation of indigenous technology for the benefit of the
developing countries in forms and in accordance with procedures which
are suited to their economies . . ." clause 4, ibid., p. 68.

(12) Ibid. p. 69.

(13) Ibid. pp. 69-84.

(14) Since petroleum has many uses, it will also have the effect of
changing the composition of petroleum demand so that defining the
shifts in demand becomes all the more difficult.

(15) There are estimates that suggest that Iran will run out of petroleum

in 20 years or so. Iran is one of the important members of OPEC.

(16) The cooperation between the developing countries and HIDC are
from two sources: cultural links and interest of Arab countries
resolve the Palestine and Jerusalem questions.

(17) See Chapter 2.

(18) India also went through a period as a police state from 1975 to
1977. There are a number of Indian scholars who tend to feel that the
logical implication of the existing social, economic, and political
structure is a police state. Egypt and Mexico, also, are no shining
examples of democracy.

(19) Thus the program of action following declaration of the establishment of NIEO contains the following specific recommendations:
"improved access to markets in developed countries through the
progressive removal of tariff and non-tariff barriers and of restrictive
business practices" (be obtained) ". . . where products of developing
countries compete with the domestic production in developed countries,
each developed country should facilitate the expansion of imports from
developing countries and provide a fair and reasonable opportunity to
the developing countries to share in the growth of the markets." UN
Monthly Chronicle, April 1974, p. 72.

(20) Ibid., p. 76.

(21) As we have explained in Part II, dependency comes from CDS. See
Chapter 7.

(22) It is suggested that Iran has purchased from the United States
armaments to the tune of $8 to 10 billion. Much of the increased
revenues from petroleum thus has ended in armaments.

(23) Up to 1976 India alone purchased, every year, food grains worth $1
billion.

(24) Many times the development policies, such as those from CDS, have
distorted the production structure of such a country and in the process
generated "dependence" on imports from the North.

(25) The market system defines prices based on the distribution of
resources (incomes). In their turn, these prices perpetuate and accentuate existing patterns of distribution. Thus, the market system favors an
existing distribution system.

(26) The price system, the military system, and the technology form
essential elements of what many scholars have called "neocolonialism."

(27) Many observers feel that it is a "wishful" document containing the

wises of the "elites" of developing countries.

(2) This shows the weaknesses of the international institutions. By concentrating on a country as a unit, this analytical tool creates biases of its own by imposing false assumptions or ignoring certain fundamental realities.

(29) It is argued in some quarters that NIEO suits the North most. Part of the reasons countries in the North favor police states in the South is that it facilitates the transfer of resources from the South to the North, since the police-state authorities are, and have to be, dependent upon the North.

(30) Some of the issues in an alternative human order are outlined in Mische and Mische (1977).

CHAPTER 16 NOTES

(1) UN Monthly Chronicle, April 1974, pp. 77-8.

(2) Organizing conferences has now become the major UN activity. Every year there are a large number of major conferences. The number of regional meetings, working group meetings, consultative bodies meetings, and so forth is numerous. These conferences pass resolutions and start working for the next conference. The list of the conferences or meetings preparatory to the Buenos Aires Conference is listed in the Report of the Buenos Aires Conference.

(3) Assume that the HTDC are fully successful and are able to completely substitute petroleum from their internal resources. This will mean that the HTDC need no imports from HIDC. The relationship between HTDC and HIDC then becomes one way and full of conflict. The situation is somewhat similar to that between the United States and Japan. Japan is having a large surplus in trade with the United States, and this has now become a major irritant in United States-Japanese relations.

(4) We are distinguishing between technologies for production and technological systems. These technological systems involve such things as supermarkets, computers, and television in the use of crowd and riot control, i.e., the use of technology in distribution and administration but not in production.

(5) One finds it everywhere. Tamils in Sri Lanka, Indians and West Indies in Great Britain, Algerians in France, Mexicans and Puerto Ricans in the United States, Turks in Germany, Greeks and Italians in Switzerland, Chinese in Malaysia and Vietnam.

(6) In November 1978, the opposition to the Shah of Iran was asking,

even threatening, that all the foreigners leave Iran. Many have left since the fall of Shah.

(7) This is the meaning of lower per capita incomes in HTDC.

(8) There are no doubt exceptions, such as imports of raw cotton in India.

(9) There is no reason to expect that the HTDC will reduce the number of mistakes ODC will have to make in adapting technology from HTDC. The nature and psychology of sales and export people ensures that such information is not passed on.

(10) We have discussed the issue of choice of technology in Chapter 5.

(11) The Tanzanian government has taken the idea of buying components of technology from HTDC seriously. Thus, it has used untied grants from the North to shop in HTDC. However, since acts are not common.

(12) P. 1 in this chapter.

(13) See Chapter 15.

(14) As we have commented earlier, dependency relations are the source and basis of exploitation. It will make the HTDC channels of exploitation of ODC by the North. Some Marxist scholars argue that some countries in HTDC, such as Brazil, are playing the role of junior partner of some developed countries, the United States in this case. Thus Brazil is considered as keeping Latin America safe for exploitation by the United States.

(15) This military muscle has been used only against neighbors. Thus, Egypt-Israel, India-Pakistan, Uganda-Tanzania have already fought wars. It has no use against the developed countries.

(16) Many times external threats of attack by the neighbor are used to keep the internal disorders and instabilities in abeyance.

(17) See Chapter 10.

(18) See Chapter 12.

(19) In a sense the whole NIEO concept is a short-term one. It deals with means and nowhere defines ends. The means imply a redistribution of economic and political power from the developed to the developing. It does not mean, imply, suggest, or ensure that this redistribution will be equitable. It is quite consistent with an inequitable redistribution.

(20) The Buenos Aires Conference on Technical Cooperation among Developing Countries passed a resolution on this subject. This resolution

urges all developing countries to "cooperate in the strengthening of their existing research and training centers with a view to providing them with a multinational scope in the framework of technical cooperation among developing countries, and to establish as necessary, new ones with the same scope."

(21) One fundamental difficulty with the international, particularly the UN, debate is that it assumes that every institution in every country is "appropriate" and should, therefore, be strengthened.

(22) Viewed from this angle, NIEO maintains and in some respects furthers neocolonialism.

(23) This will involve a different orientation of these UN bodies, from a hierarchical to a more convivial one.

(24) In the principle of trusteeship we have suggested a principle for the formation of new institutions. See Chapter 12.

Bibliography

Abramovitz, Moses (1962). "Economic Growth in the United States: A Review Article." American Economic Review, LII: 762-82.
_____ andDavid, Paul A. (1973b). "Economic Growth in America: Historical Parables and Realities." De Economist, Vol. 121: 3.
_____ (1973a) "Reinterpreting Economic Growth: Parables and Realities of the American Experience." American Economic Review, vol. 58, no. 2.
Acharya, Shankar N. (1974) Fiscal/Financial Intervention, Factor Prices and Factor Proportions: A Review of Issues. IBRD Bank Staff Working Paper No. 183. Washington, D.C.: International Bank for Reconstruction and Development.
Achilladelis, Basil (1974) Emerging Changes in the Petrochemical Industry: An Overview. OECD Development Center, Occasional Paper No. 3. Paris: Organization for Economic Cooperation and Development.
Adelman, Irma (1975) "Development Economics: A Reassessment of Goals." American Economic Review, Papers and Proceedings 65: 302-09.
_____ (1963) "An Econometric Analysis of Population Growth." American Economic Review, 53: 314-39.
_____ (1961) Theories of Economic Growth and Development. Palo Alto, CA: Stanford University Press.
_____ andMorris, Cynthia, T. (1973) Economic Growth and Equity in Developing Countries. Palo Alto, CA: Stanford University Press.
Agarwala, A.N. and Singh, S.P., eds. (1958) The Economics of Development. Oxford, England: Oxford University Press.
Agency for International Development (1962) Science, Technology and Development. 12 Vols. U.S. papers presented for the UN Conference on the Application of Science and Technology for the Benefit of the Less Developed Areas. Washington, D.C.: U.S. Government Printing Office.
Aggarwal, Partap C. (1973) The Green Revolution and Rural Labor, Delhi: Sri Ram Centre.

Ahmad, S. (1966) "On the Theory of Induced Innovation." Economic Journal vol. LXXVI.

Alchian, Armen (1963) "Reliability of Progress Curves in Airframe Production." Econometrica, vol. XXXI, no. 4.

Alker, Harward R. Jr.; Bloomfield, Lincoln P.; and Choucri, Nazli (1974) Analyzing Global Interdependence. Center for International Studies, Cambridge, MA: M.I.T.

Allal, Moise (1974) Appropriate Technology vol. 1. London.

_____ (1975) Selection of Road Projects and the Identification of the Appropriate Road Construction Technology: General Considerations. World Employment Programme Research, WEP-22/WP14. Geneva: International Labor Office.

Alonzo, William (1968) "Urban and Regional Imbalances in Economic Development." Economic Development and Cultural Change 17: 1-14.

American Friends Service Committee (1977) Taking Charge. New York: Bantam.

Amin, Samir (1974) Accumulation on a World Scale. 2 Vols. New York: Monthly Review Press.

_____ (1977) Imperialism and Unequal Development. New York: Monthly Review Press.

_____ (1975) Neo-Colonialism in West Africa. New York: Monthly Review Press.

_____ (1976) Unequal Development: An Essay on the Social Formations on Peripheral Capitalism. New York: Monthly Review Press.

Apter, David E. and Goodman, Louis Wolf, eds. (1976) The Multinational Corporation and Social Change. New York: Praeger Publishers, Inc.

Arendt, Hannah (1969) The Origins of Totalitarianism. Cleveland: World Publishing Company.

Arrow, Kenneth J. (1969) "Classificatory Notes on the Production and Transmission of Technological Knowledge." American Economic Review vol. LIX, no. 2.

_____ (1962a) "The Economic Implications of Learning by Doing." Review of Economic Studies, vol. XXIX: 2.

_____ (1962b) "Economic Welfare and the Allocation of Resources for Invention." The Rate and Direction of Invention Activity, Universities-National Bureau Committee for Economic Research. Princeton, NJ: Princeton University Press.

Arubuthnot, John (1773) An Inquiry into the Connection Between the Present Price of Provisions and the Size of Farms, by a Farmer. London.

Asher, Ephrain (1972) "Industrial Efficiency and Biased Technical Change in American and British Manufacturing: The Case of Textiles in the Nineteenth Century." Journal of Economic History vol. 32, no. 2.

Asher, R.E.; Hagen, E.E.; Hirschman, A.O.; Colm, G.; Geiger, T.; Mosher, A.T.; Echaus, R.S.; Borman, M.J.; Anderson, C.A.; and Wriggins, H. (1972) Development of Emerging Countries: An Agenda for Research. Washington, D.C.: The Brookings Institution.

Ashton, T.S. (1955) An Economic History of England: The Eighteenth Century. London: Methuen and Co., Ltd.

Asimakopulos, A. (1963) "The Definition of Neutral Inventions."
Economic Journal, LXXIII: 674-80.
_____ andWeldon, J.C. (1963b) "The Classification of Technical
Progress in Models of Economic Growth." Economica, XXX: 372-86.
_____ (1965) "A Synoptic View of Some Simple Models of Growth."
Canadian Journal of Economics and Political Science, XXXVI: 52-79.
_____ (1963a) "Sir Roy Harrod's Equation of Supply." Oxford Economic
Papers XV: 266-72.
Atkinson, Anthony B. (1975) The Economics of Inequality. London:
Oxford University Print.
_____ andStiglitz, Joseph E. (1969) "A New View of Technological
Change." Economic Journal, Vol. LXXIX: 315.
Aubrey, Henry G. (1955) "Investment Decisions in Underdeveloped
Countries." In Capital Formation and Economic Growth. National
Bureau of Economic Research.
_____ (1951) "Small Industry in Economic Development." Social
Research 18: 269-312.
Auciello, Kay Ellen (1976) Bibliography of Intermediate Technology
Materials Held at the International Development Data Center.
Atlanta: Engineering Experiment Station, Georgia Institute of
Technology.
Aurobindo, Sri (1974) The Future Evolution of Man. Wheaton, Il:
Theosophical Publishing House.
Backstrand, Goran and Ingelstam, Lars (1977) "Should We Put Limits on
Consumption?" The Futurist, June, pp. 157-62.
Baer, W., and Herne, M. (1966) "Employment and Industrialization in
Developing Countries." Quarterly Journal of Economics, Vol. LXXX.
Baeumer, Ludwig (1972) Importance of Industrial Property Protection in
Developing Countries. (ID/WG.1304/4). Vienna: United Nations Indus-
trial Development Organization.
Baines, E. (1835) A History of the Cotton Manufacture in Great Britain.
London, 1835.
Balassa, Bala (1970) "Growth Strategies in Semi-Industrial Countries."
Quarterly Journal of Economics, Vol. 84, No. 1: 24-47.
_____ (1961) "Toward a Theory of Economic Integration." Kyklos, Vol.
14: 1-14.
Balasubramanyam, V.N. (1973) International Transfer of Technology to
India. New York: Praeger Publishers.
Baldwin, J. and Brand, Stewart, eds. (1978) Soft-Tech. Baltimore:
Penguin Books.
Baldwin, Robert E. (1969) "The Case Against Infant-Industry Tariff
Protection." Journal of Political Economy vol. LXXVII, no. 3.
_____ (1966) Economic Development and Growth. New York: John
Wiley and Sons.
Balis, John S. (1976) Appropriate Technology for Agricultural Develop-
ment. Washington, D.C.: U.S. Agency for International Develop-
ment.
Balogh, Thomas (1961) "Economic Policy and the Price System."
Economic Bulletin for Latin America.

_____ (1974) The Economics of Poverty. White Plains, NY: M.E. Sharpe, Inc.

Baranson, Jack (1963) "Economic and Social Considerations in Adapting Technologies forDeveloping Countries." Technology and Culture. Vol. IV: 22-29.

_____ (1962a) "Implementing Technology Programs for Underdeveloped Countries." Oregon Business Review XXVI: 1-4.

_____ (1969) Industrial Technologies for Developing Economies. New York: Praeger Publishers.

_____ (1960) "National Programs for Science and Technology in the Underdeveloped Areas." Bulletin of the Atomic Scientists vol. XVI: 151-54.

_____ (1974) "Technical Improvements in Developing Countries." Finance and Development, 11:2-5.

_____ (1962b) "Technological Opportunities for Underdeveloped Economies." International Development Review IV: 24-27.

_____ (1967) Technology for Undeveloped Areas: An Annotated Bibliography. Oxford, England: Pergamon Press.

_____ (1970) "Technology Transfer through the International Firm." American Economic Review Papers and Proceedings 60: 435-40.

Bardhan, Pranab K. (1975) Agricultural Development and Land Tenancy in a Peasant Economy: A Theoretical and Empirical Analysis. Ottawa, Canada: International Development Research Center, Income Distribution Division.

_____ andSrinivasan, T.N. (1971) "Cropsharing Tenancy in Agriculture." American Economic Review 61: 48-64.

Barnet, Richard J. and Muller, Ivan. 1974. Global Reach. New York: Simon and Schuster.

Bar-Zakay, S. (1972) "Technology Transfer Model." Industrial Research and Development News, 6:1-11.

Basche, James R., Jr. and Duerr, Michael G. (1975) International Transfer of Technology: A Worldwide Survey of Chief Executives. New York: Conference Board, Inc.

Battelle Memorial Institute (1978) An Analysis of Federal Incentives Used to Stimulate Energy Production. Washington, D.C.: G.P.O.

Bauer, Peter T. (1959) "International Economic Development." Economic Journal 69: 105-23.

_____ andYamey, Basil S. (1951) "Economic Progress and Occupational Distribution." Economic Journal vol. 61: 741-55.

_____ (1957) The Economics of Underdeveloped Countries. Chicago: University of Chicago Press.

Beckerman, W. and Bacon, R. (1960) "International Comparison of Income Levels: A Suggested New Measure." Economic Journal 76: 518-36.

Behrman, J.N. (1962) "Foreign Investment and the Transfer of Knowledge and Skills." U.S. Private and Government Investment Raymond Michesell, ed. Eugene,OR: University of Oregon Press, 114-36.

Bell, Clive (1972) "The Acquisition of Agricultural Technology: Its Determinants and Effects." The Journal of Development Studies 9: 123-60.

Bell, Thomas George (1856) "A Report Upon the Agriculture of the Country of Durham." Journal of the Royal Agricultural Society of England, vol. XVII.

Bender, Tom (1975) Sharing Smaller Pies, Salem, OR: Tom Bender.

Bennett, Merrill K. (1951) "International Disparities in Consumption." American Economic Review 41: 632-49.

Benveniste, Gay and Moran, William E., Jr. (1962) Handbook of African Economic Development. New York: Praeger Publishers.

Bereano, Philip L. (1976) "Alternative Technology: Is Less More?" Science for the People, September-October.

Beres, Louis Rene and Targ, Harry R. (1974) Reordering the Planet: Constructing Alternative World Futures. Boston, MA: Allyn and Bacon.

Berger, Suzanne (1974) "The Uses of the Traditional Sector: Why the Declining Classes Survive." Fabio Luca Cavazza and Stephen R. Graubard, In Il Caso Italiano. Milan, Italy: Garzanti.

Bernstein, Scott (1978) "Small-scale Sewage Options Gain in Chicago." Self-Reliance, July-August, pp. l-ll.

Berril, K., ed. (1964) Economic Development with Special References to East Asia. New York: St. Martin's Press.

Berry, A. (1972) Unemployment as a Social Problem in Urban Columbia: Some Preliminary Hypotheses and Interpretation. Yale University, Economic Growth Center, Discussion Paper No. 145. New Haven, CT: Yale University Press.

Berry, Wendell (1977) The Unsettling of America: Culture and Agriculture. San Francisco, CA: Sierra Club Books.

Best, R.H. and Coppock, J.T. (1962) The Changing Use of Land in Britain. London: Faber and Faber.

Bettelheim, C. (1976) Economic Calculation and Forms of Property. New York: Monthly Review Press.

_____ (1978) "The Great Leap Backward." In China Since Mao. Monthly Review, special edition, July-August 1978, 37-130.

_____ (1968) India Independent. New York: Monthly Review Press.

Bhagwati, Jagdish, ed. (1970) Amount and Sharing of Aid. Washington, D.C.: Overseas Development Council.

_____ (1972) Economics and World Order: From the 1970s to the 1980s. New York: Free Press.

_____ (1967) "The Pure Theory of International Trade: A Survey." In Surveys in Economic Theory II. American Economic Association – Royal Economic Society. New York: St. Martin's Press.

_____ andChakravarty, S. (1969) "Surveys of National Economic Policy: Issues and Policy Research." The American Economic Review, Supplement 59: 118.

_____ andDesai, P. (1970) India: Planning for Industrialization. Oxford: Oxford University Press.

Bhalla, A.S., ed. (1964) "Investment Allocation and Technological Choice: A Choice of Cotton Spinning Techniques." Economic Journal, vol. LXXIV.

_____ (1975) Technology and Employment in Industry. Geneva: International Labor Office.

Bhalla, A.S. and Gaude, J. (1973) Appropriate Technologies in Services with Special Reference to Retailing. Geneva: International Labor Office, mimeo.

_____ andStewart, F. (1976) "International Action for Appropriate Technology." Background paper for the World Employment Conference. Geneva: International Labor Office.

Bhargava, Ashok (1975) "A Critical Look at India's Development Strategy." R. Diwan, ed. Paper presented at the 1st Conference of the Association for Indian Economic Studies, Albany, N.Y.

Bhatia, Ramesh (1974) "The Oil Crisis: An Economic Analysis and Policy Imperatives." Economic and Political Weekly, July 27, 1191-1203.

Bidwell, Percy W. and Falconer, John I. (1925) History of Agriculture in the Northern United States. (Publication No. 358) Washington, D.C.: Carnegie Institute of Washington.

Binder, L.; Coleman, J.S.,; La Palombara, J.; Pye, L.W.; Verba, S.: and Weinter, M. (1971) Crises and Sequences in Political Development. Princeton, NJ: Princeton University Press.

Black, Eugene R. (1963) The Diplomacy of Economic Development and Other Papers. New York: Atheneum.

Blaugh, M. (1974) Education and Employment Problems in Developing Countries. Geneva: International Labor Office.

Blitzer, C.R.; Clark, P.B.; and Taylor, L. (1975) Economy-Wide Models and Development Planning. Oxford, England: Oxford University Press.

Bloom, G.F. "Union Wage Pressure and Technological Discovery." American Economic Review, vol. XLI, September 1951.

Bogue, Allan G. (1963) "Farming in the Prairie Peninsula, 1830-1880." Journal of Economic History, vol. 23.

Bohm-Bawerk, E. von (1959) History and Critique of Interest Theories, vol. I: Capital and Interest. Translated by G.D. Huncke and H.F. Sennholz. South Holland, IL: Libertarian Press.

_____ Positive Theory of Capital, vol. II: Capital and Interest (see above).

_____ Further Essays on Capital and Interest, vol. III: Capital and Interest (see above).

_____ Economic Choice of Human and Physical Factors in Production: Contributions to Economic Analysis. Amsterdam: North Holland Publishing Co., 1964.

Borgese, G.A. (1953) Foundations of the World Republic, Chicago.

Boulding, Kenneth (1968) Beyond Economics: Essays on Society, Religion, and Ethics, Ann Arbor, MI: University of Michigan Press.

_____ (1970) Economics as a Science New York: McGraw-Hill.

_____ (1973) Economics of Love and Fear. Belmont, CA: Wadsworth Publishing Co.

_____ (1955) "In Defense of Statics." Quarterly Journal of Economics 69: 485-502.

_____ (1949-50) "Income or Welfare." Review of Economic Studies

_____ et al. (1977) Social System of the Planet Earth.

_____ (1962) "The Relations of Economics, Political, and Social Systems." Social and Economic Studies, vol. II: 351-62.

Bowley, A.L. (1898) "Statistics of Wages in the United Kingdom During the Past Hundred Years." Journal of the Royal Statistical Society, vol. LXI.

Boyle, Godfrey; Harper, Peter; and the editors of Undercurrents (1976) Radical Technology: Food and Shelter, Tools and Materials, Energy and Communications, Autonomy and Community. New York: Pantheon Books.

Branfman, Fred and LaMar, Steve (1978) Jobs from the Sun. Los Angeles, CA: California Public Policy Center.

Brannon,Gerard (1977) Energy Taxes and Subsidies. Cambridge, MA: Ballinger Publishing Co.

Brewer, Gary D. and Brunner, Ronald D., eds. (1975) Political Development and Change. New York: The Free Press.

Bronfenbrenner, Martin (1955) "The Appeal of Confiscation in Economic Development." Economic Development and Cultural Change 3: 201-18.

_____ (1963) "Second Thoughts of Confiscation." Economic Development and Cultural Change, 2: 367-71.

Brooks, Harvey (1967) "National Science Policy and Technology Transfer." Proceedings of a Conference on Technology Transfer and Innovation, May 15-17, 1966, sponsored by the National Planning Association and National Science Foundation (NSF 65-7). Washington, D.C.: U.S. Government Printing Office, 53-64.

Brown, Lester R. (1974) In the Human Interest. New York: W.W. Norton.

_____ (1978) The Twenty-Ninth Day: Accommodating Human Needs and Numbers to the Earth's Resources. New York: W.W. Norton.

Brown, M. (1966) On the Theory and Measurement of Technological Change. Cambridge: Cambridge University Press.

Brown, Norman L., ed. (1977) Renewable Energy Resources and Rural Applications in the Developing World. Boulder, CO: Westview Press.

_____ andHowe, James W. "Solar Energy for Village Development." Science, February 10, 1978, 651-7.

Brown, Seyom (1974) New Forces in World Politics. Washington, D.C.: The Brookings Institution.

Bruni, L. "Internal Economics of Scale with a Given Technique." Journal of Industrial Economics, Vol. 12, July 1964.

Bruton, Henry J. (1960) "Contemporary Theorizing on Economic Growth." In Theories of Economic Growth. Edited by Bert F. Hoselitz. New York: Free Press.

_____ (1965) Principles of Development Economics, Englewood Cliffs, NJ: Prentice Hall.

Bulfin, and Greenwell (1977) Appropriate Technology in Less Developed Countries. Tucson: University of Arizona.

Burke, Fred G. 1965. Tanganyika: Preplanning. Syracuse, NY: Syracuse University Press.

Burmeister, Edwin and Dobell, Rodney (1971) Mathematical Theories of Economic Growth. New York: Macmillan.

Burns, Scott (1975) Home, Inc. New York: Doubleday and Co.

Caincross, A.K. (1962) Factors in Economic Development. London: George Allen & Unwin Ltd.

Caird, James (1878) The Landed Interest and the Supply of Food. London.

Calvert, Monte A. (1967) The Mechanical Engineer in America, 1830-1910. Baltimore, MD: The Johns Hopkins University Press.

Cameron, Rondo (1967) "Economic Development: Some Lessons of History for Developing Nations." American Economic Review 57: 312-24.

Canadian Hunger Foundation (1976) A Handbook of Appropriate Technology. Ottawa: Canadian Hunger Foundation.

Caporaso, James A., ed. (1978) "Dependence and Dependency in the Global System." International Organization, Winter.

Carr, M., ed. (1976) Economically Appropriate Technologies for Developing Countries: An Annotated Bibliography. Intermediate Technology Department Group. Forest Grove, OR: International Scholarly Book Service.

Caves, R.E. and Jones, R.W. (1973) World Trade and Payments.

Center for Science in the Public Interest (1977) 99 Ways to a Simple Lifestyle. Bloomington, IN: Indiana University Press.

Chamber of Commerce of the United States (1961) "What is Economic Growth?" In The Promise of Economic Growth. Washington, D.C.

Chambers, J.D. and Mingay, G.E. (1966) The Agricultural Revolution 1750-1880 London: B.T. Batsford.

Chenery, H.B. (1961) "Comparative Advantages and Development Policy." American Economic Review 51: 18-51.

_____; Ahluwalia, Montek, S.; Bell, C.L.G.; Duloy, John H.; and Jolly, Richard (1974) Redistribution with Growth. Oxford, England: Oxford University Press.

_____ andBruno, M. (1962) "Development Alternatives in an Open Economy." Economic Journal 72: 79-103.

_____ andStrout, A. (1966) "Foreign Assistance and Economic Development." American Economic Review, 679-733.

_____ andSyrquin, Moises (1975) Patterns of Development 1950-1970. London: Oxford University Press for the World Bank.

_____ andTaylor, Lance (1968) "Developing Patterns: Among Countries and Over Time." Review of Economics and Statistics 50: 391-416.

Choi, Harry Y.H. (1973) The RANN Program: Potential Benefits to Developing Countries. U.S. Agency for International Development, Office of Science and Technology.

Choucri, Nazli (1976) Technological Choice and Political Development. Cambridge, MA: Massachusetts Institute of Technology, Department of Political Science.

Clark, Colin (1957) Conditions of Economic Progress, 3rd ed. New York: Macmillan.

_____ (1965) "The Fundamental Problems of Economic Growth." Weltwirtschlaftliches Archieve 94, no. 1: 1-9.

_____ (1973) Value of Agricultural Land. New York: Pergamon Press.

Clark, V.S. (1929) History of Manufactures in the United States vol. I. New York: McGraw-Hill.

Clapham, J.H. (1952) An Economic History of Modern Britain. Cambridge: Cambridge University Press.

Clarke, Robin (1977) Building for Self-Sufficiency. New York: Universe Books.
Coale, A.J. and Hoover, E.M. (1958) Population Growth and Economic Development in Low Income Countries. Princeton, NJ: Princeton University Press.
Cochrane, Willard W. (1969) The World Food Problem: A Guardedly Optimistic View. New York: Cromwell.
Cockcroft, John (1966) Technology for Developing Countries. London: Overseas Development Institute.
Cohen, Allan R. (1974) Tradition, Change and Conflict in Indian Family Business. The Hague: Mouton Publishers.
Collins, E.J.T. (1969b) "Harvest Technology and Labor Supply in Britain, 1790-1870." Economic History Review, 2nd Series, vol. XXII, no. 3.
_____ (1969a) "Labor Supply and Demand in European Agriculture 1800-1880." In E.L. Jones and S.J. Woolf, eds. Agrarian Change and Economic Development: The Historical Problems. London: Methuen and Co., Ltd.
Commager, Henry Steele (1966) Freedom and Order. New York: George Braziller.
Commoner, Barry (1971) The Closing Circle: Nature, Man and Technology, New York: Alfred A. Knopf.
_____ (1976) Poverty of Power. New York: Alfred A. Knopf.
_____ et al. (1975) A Comparison of the Production, Economic Returns, and Energy Intensiveness of Corn Belt Farms That Do and Do Not Use Inorganic Fertilizers and Pesticides. St. Louis, MO: Washington University.
Congdon, R.J., ed. (1977) Introduction to Appropriate Technology. Emmaus, PA: Rodale Press.
Cooper, Charles, 1972. Science Policy and Technological Change in Underdeveloped Economies, World Development.
_____ (1972) "Science, Technology and Production in the Underdeveloped Countries: An Introduction." The Journal of Development Studies 9: 1-18.
Courtney, William H. and Leipziger, Danny M. (1974) Multinational Corporations in LDCs: The Choice of Technology. Bureau for Program and Policy Coordination, Discussion Paper, No. 29. U.S. Agency for International Development, Washington, D.C.
Cousins, Norman (1961) In Place of Folly. New York: Harper and Row Publishers.
Crane, Diana (1974) An Inter-Organizational Approach to the Development of Indigenous Technological Capabilities: Some Reflections on the Literature. Organization for Economic Cooperation and Development, Development Center, Occasional Paper, No. 3. Paris: Organization for Economic Cooperation and Development.
Curries, Lauchlin (1966) Accelerating Development: The Necessity and the Means. New York: McGraw-Hill.
Dag Hammarskjold Foundation (1975) Another Development.
Dahl, Norman C. "Absorption of Technology in Developing Countries." Paper presented at Symposium on Technology, Modernization, and Cultural Impact, May 8-9, 1974 at Iowa State University.

Dahlberg, K. (1978) "An Evaluation of Research Strategies for Developing Appropriate Agricultural Systems and Technologies." Paper presented to International Studies Association.

Daly, Herman (1977) Steady-State Economics: The Economics of Biophysical Equilibrium and Moral Growth. San Francisco, CA: W.H. Freeman.

_____ ed. (1972) Towards a Steady State Economy. San Francisco, CA: W.H. Freeman.

Darrow, Ken and Pam, Rick (1976) Appropriate Technology Sourcebook. Stanford, CA: Volunteers in Asia.

Dasgupta, Partha; Sen, Amartyn; and Marglin, Stephen (1972) Guidelines for Project Evaluation. Project Formulation and Evaluation Series, No. 2 (ID/SER.H/2). Vienna: United Nations Industrial Development Organization.

Datt, Ruddar and Sundharam, P.M. (1972) India Economy, 8th ed. N. Dillon, S. Chand and Co.

David, Paul A. (1969) "A Contribution to the Theory of Diffusion." Stanford Center for Research in Economic Growth, Memorandum No. 71.

_____ (1970b) "Labor Productivity in English Agriculture, 1850-1914: Some Quantitative Evidence on Regional Differences." Economic History Review, 2nd Series, vol. XXIII, no. 3.

_____ (1971) "The Landscape and the Machine: Technical Interrelatedness, Land Tenur, and the Mechanization of the Victorian Corn Harvest." In D.N. McCloskey, ed., Essays on a Mature Economy: Britain after 1840. London: Methuen and Co., Ltd.

_____ (1970a) "Learning by Doing and Tariff Protection: A Reconsideration of the Case of the Ante-Bellum United States Cotton Textile Industry." Journal of Economic History, vol. XXX, no. 3.

_____ (1966b) "Measuring Real Net Output: A Proposed Index." Review of Economics and Statistics XLVIII: 419-25.

_____ (1966a) "The Mechanization of Reaping in the Ante-Bellum Midwest." In H. Rosovsky, ed. Industrialization in Two Systems. New York: John Wiley & Sons.

_____ andvan de Klundert, T. 1965. "Biased Efficiency Growth and Capital-Labor Substitution in the U.S., 1899-1960." American Economic Review vol. 53, no. 3.

_____ (1962) "The Deflation of Value Added." Review of Economics and Statistics XLIV: 148-55.

Davis, Lance E., et al. (1972) American Economic Growth New York: Harper & Row.

Deane, Phillis and Cole, W.A. (1962) British Economic Growth, 1668-1955. Cambridge: Cambridge University Press.

_____ andMitchell, B.R. (1962) Abstract of British Historical Statistics. Cambridge: Cambridge University Press.

Denison, Edward T. (1962) The Sources of Economic Growth in the United States. New York: Committee for Economic Development.

_____ (1967) Why Growth Rates Differ. Washington, D.C.: The Brookings Institution.

Denison, E.F. "Some Major Issues in Productivity Analysis: An Examination of Estimates by Jorgensen and Griliches." Survey of Current Business, Vol. XLIX, May 1969, 1-30.

_____ with Poullier, J.P. (1967) Why Growth Rates Differ Washington, D.C.: The Brookings Institution.

_____ (1957) "Theoretical Aspects of Quality Change, Capital Consumption, and Net Capital Formation." Problems of Capital Formation: Concepts, Measurement, and Controlling Factors, Studies in Income and Wealth, vol. 19, Princeton, NJ: Princeton University Press.

DeRoop, Robert S. (1975) Eco-Tech: The World-Earther's Guide to the Alternative Society. New York: Delacorte.

Desai, Suresh A. (1976) "Wither India? A Gandhian Alternative." In Romesh Diwan, ed. Issues in Indian Economy, 33-45.

Design Alternatives (1978) Workshop on Appropriate Technology: Final Report and Proceedings. Washington, D.C.: U.S. National Science Foundation.

De Varies, Eghert (1964) "International Transfers of Knowledge and Capital." Natural Resources and International Development, Edited by Marion Clawson. Baltimore, MD: Johns Hopkins University Press, 415-35.

Dewey, D. (1965) Modern Capital Theory. New York: Columbia University Press.

Dhar, P.N. and Lydall, H.F. (1961) The Role of Small Enterprises in Indian Economic Development. New York: Asia Publishing House.

Diamond, Peter A. (1965a) "Disembodied Technical Change in a Two-Sector Model." Review of Economic Studies XXXII: 161-68.

_____ (1965b) "Technical Change and the Measurement of Capital and Output." Review of Economic Studies XXXII: 289-98.

_____ andMirrless, James A. (1971) "Optimal Taxation and Public Production." American Economic Review LXI: 261-78.

Diaz, Alejandro C.F. (1970) Essays on the Economic History of the Argentine Republic.

_____ (1966) Exchange-Rate Devaluation in a Semi-Industrial Country: The Experience of Argentina 1961-1966. Cambridge, MA: MIT Press.

_____ (1965) "Industrialization and Labor Productivity Differentials." Review of Economics and Statistics, vol. XLVII.

Dickson, David (1975) The Policies of Alternative Technology. New York: University Books.

Diwan, Romesh K. (1975) "Agriculture in India: Problems and Prospects." In J. Uppal, ed. India's Economic Problems. New York: McGraw-Hill, pp. 45-63.

_____ (1974b) "Another Look at the Green Revolution." Paper presented at the 1974 meeting of the New York State Conference on Asian Studies at Cornell University, Ithaca, New York.

_____ (1967) "Are we Saving Enough for Our Plans? Economic and Political Weekly 2: 293-302.

_____ (1966) "The Debate on Food in India: Some Relevant Variables." Economic and Political Weekly 1: 849-56.

_____ (1977a) "Development, Education and the Poor: Context of South

Asia" Economic and Political Weekly, February 26, 1977, 401-8.
_____ (1978a) "Energy Implications of Indian Economic Development: Decade of 1960-70 and After." Journal of Energy and Development. Spring, 318-338.
_____ (1974c) "Gandi and Modernization of Modes and Production." Paper presented at the Conference on South Asia, Oshkosh, WI, November 15-16, 1974.
_____ ed. (1976b) Issues in Indian Economy. Papers and Proceedings of the 1st Conference of Association of Indian Economics Studies.
_____ (1977c) "1977 Elections in India: An Interpretation." Journal of Asian Thought and Society. September, pp. 262-264.
_____ (1971) "Planning for the Poor." Economic and Political Weekly Vol. 4, no. 4: pp. 529-37.
_____ (1974a) "Plannostructure and Garigi Hatao." Paper presented at the India Forum, Chicago, February 10, 1974 and at the South Asia Seminar, University of Chicago, February 12, 1974.
_____ (1973b) "Plannostructure and Poverty in India." Paper presented at the New York State Conference on Asian Studies, New Paltz, NY.
_____ (1977b) "Small is Beautiful in India." Asian Thought and Society. September, pp. 196-206.
_____ (1976a) "Some Elements in Gandhian Theory of Development." Paper given at CASAS 1976 Meeting in Quebec City, Quebec. To appear in its Proceedings.
_____ (1968b) "A Test of the Two Gap Theory of Economic Development." Journal of Development Studies.
_____ (1973a) "Trade Between Unequal Partners." Economic and Political Weekly VII: 213-20.
_____ (1977d) "Unemployment and Employment: Some Conceptual Issues." Mimeo.
_____ andAlavi, Kamelia (1978c) "Technical Dependence and Economic Development in India: An Empirical Analysis." Paper Presented in ASSA Conference, August 29-31, Chicago, Illinois.
_____ andBhargava, Ashok (1978b) "Janata Government: Economic Policies, Performances, Problems and Proposals." Paper given at Association of Asian Studies Meeting in Chicago, March 30-April 2.
_____ andGidwani, Sushila (1978d) "Elements in Ghandian Economics." Paper presented at the Association of Indian Economics Studies Sessions in the 1978 Annual Meeting of the Allied Social Sciences Associations, Chicago, August 28-31.
_____ andMarwah, Kanta (1976) "Transfer from Poor to Rich Countries: An Analysis of World Exports." Economic and Political Weekly, February.
_____ andGujarati, Damodar N. (1968a) "Employment and Productivity in Indian Industries: Some Questions of Theory and Policy." Artha Vijnana, Journal of the Gokblc Institute of Politics and Economics, India, vol. 10, pp. 29-67.
Dolman, A.J. and Ettinger, Jan Van, eds. (1978) Partners in Tomorrow: Strategies for a New International Order. New York: E.P. Dutton.
Domar, E.D. (1964) "Economic Growth and Productivity in the U.S., Canada, U.K., Germany and Japan in the Post-War Period." Review of Economics and Statistics 46: 33-40.

_____ (1957) Essays in the Theory of Economic Growth. Fair-lawn: Oxford University Press.

_____ "On Total Productivity and All That." Journal of Political Economy, Vol. LXX, December 1962, 597-608.

_____ "On the Measurement of Technological Change." Economic Journal, Vol. LXXI, December 1961, 709-29.

_____ "Total Productivity and the Quality of Capital." Journal of Political Economy, Vol. LXII, December 1963, 586-8.

Dorf, Richard D. and Hunger, Yvonne, eds. (1978) Appropriate Visions: Technology, the Environment, and the Individual. San Francisco, CA: Boyd & Fraser.

Dorrance, Grace S. "The Effect of Inflation on Economic Development." International Monetary Staff Papers, Vol. 10, March 1963, 1-47.

Dovring, F. (1965) "The Transformation of European Agriculture." Cambridge Economic History of Europe. Edited by M.M. Postan and J.H. Habakkuk, Vol. 6, Part II. Cambridge: Cambridge University Press.

Dricker, Peter F. "The Technological Revolution, Notes on the Relationship of Technology, Science and Culture." Technology and Culture, Vol. II, No. 4, Fall 1961, 342-51.

Dubos, Rene (1974) Beast or Angel? New York: Charles Scribner's Sons.

Dunkerley, Harold B. "The Choice of Appropriate Technologies." Finance and Development, Vol. 14, No. 3, September 1977, 36-9.

Dutt, Romesh C. (1902) History of India. 2 Volumes. Oxford: Oxford University Press.

East-West Center, Technology and Development Institute. Pilot Study on the Generation and Diffusion of Adaptive Technology in Indonesia: A Summary Report, Honolulu, Hawaii.

Eckaus, Richard S. (1977) "Appropriate Technologies for Developing Countries." Washington, D.C.: National Academy of Science.

_____ (1962b) "Education and Economic Growth." In Economics of Higher Education. Edited by S.J. Mushkin. Washington, D.C.: U.S. Department of Health, Education and Welfare, 102-28.

_____ "The Factor Proportions Problem in Underdeveloped Areas." American Economic Review, Vol. 45, September 1955, 539-65.

_____ "Notes on Invention and Innovation in Less Developed Countries." American Economic Review, Vol. 56, May 1966, 98-109.

_____ (1962a) "Technological Change in Less Developed Areas." In Development of the Emerging Countries. Edited by Robert E. Asher. Washington, D.C.: The Brookings Institution.

_____ andParikh, Kriti S. (1968) Planning for Growth: Multisectoral, Intertemporal Models Applied to India. Cambridge, MA: MIT Press.

Eckstein, Otto "Investment Criteria for Economic Development and the Theory of Intertemporal Welfare Economics." Quarterly Journal of Economics, Vol. 71, February 1957, 56-85.

Economic Commission for Latin America (ECLA) (1966) Choice of Techniques in the Latin American Textile Industry.

Economic Justice (1975) A Special Issue of New Catholic World. September/ October.

Edwards, Edgar O. (1974) Employment in Developing Nations. New York: Columbia University Press.

Eibl-Eibesfeldt, Irenaua (1971) Love and Hate: The Natural History of Behavior Problems. New York: Holt, Rinehart and Winston, Inc.

Elgin, Duane S. and Mitchell, Arnold "Voluntary Simplicity: Life-Style of the Future?" The Futurist, August 1977.

Ellis, Howard S. "How Culture Shapes Economic Growth." Arizona Review, Vol. 20, No. 1, January 1971, 1-9.

_____ andBuchanan, Norman (1955) Approaches to Economic Development. New York: Twentieth Century Fund.

Ellis, William N. "AT: The Quiet Revolution." Bulletin of the Atomic Scientists, November 1977.

Ellul, Jacques (1964) The Technological Society. New York: Alfred A. Knopf.

Elmandjra, M. UNESCO Information Exchange and Development: Background Paper for DEVSIS Feasibility Study for the Preliminary Design for an International Information System for the Development Sciences (DEVSIS/FS/ME/5), March 3, 1975. Paris: United Nations Educational, Scientific, and Cultural Organization, 1975.

Enke, Stephen "Economists and Development: Rediscovering Old Truths." Journal of Economic Literature, Vol. 7, December 1969, 1125-39.

Enos, J.L. (1976) More (or Less) on the Choice of Techniques, With a Contemporary Example. Magdalon: Oxford Mimeo.

Erb, Guy F. and Kallab, Valeriana (1975) Power and Interdependence: World Politics in Transition. Boston, MA: Little, Brown & Co.

Erikson, Erik H. (1969) Gandhi's Truth. New York: W.W. Norton & Co.

Evans, G.E. (1960) The Horse in the Furrow. London: Faber and Faber.

Evenson, Robert E. and Kislev, Yoav (1975) Agricultural Research and Productivity. New Haven, CT: Yale University Press.

Falcon, Walter "The Green Revolution: Generations of Problems." The American Journal of Agricultural Economics, Vol. 52, December 1970, 698-712.

Falk, Richard (1966) The Strategy of World Order. New York: Institute for World Order.

_____ (1975) A Study of Future Worlds. New York: The Free Press.

_____ (1975) This Endangered Planet. New York: The Free Press.

Falkus, Malcolm E., ed. (1968) Readings in the History of Economic Growth, Fairlawn: Oxford University Press.

Fanon, Frantz (1968) Black Skin, White Masks. New York: Grove Press.

_____ (1966) The Wretched of the Earth. New York: Grove Press.

Fei, J.C.H. and Ranis, Gustav (1974) Growth and Employment in South Korea and Taiwan. New Haven, CT: Yale University Press.

_____ (1964) Development of the Labor Surplus Economy. Homewood, IL: Richard D. Irwin.

Feinstein, Otto, ed. (1964) Two Worlds of Change: Readings in Economic Development, Anchor Books.

Felix, D. (1966) "Beyond Import Substitution: A Latin American Dilemma." Mimeo.

Fellner, William (1962) "Does the Market Direct the Relative Factor-Saving Effects of Technological Progress?" In The Rate and Direction of Inventive Activity. Universities-National Bureau Committee for Economic Research. Princeton, NJ: Princeton University Press.

_____ (1969) "Specific Interpretations of Learning by Doing." Journal of Economic Theory, Vol. I, 2.

_____ "Two Propositions in the Theory of Induced Innovations." Economic Journal, Vol. LXXI, June 1961.

Ferguson, C.E. (1969) The Neo-Classical Theory of Production and Distribution. Cambridge: Cambridge University Press.

Figlewski, Stephen (1975) The Burden of Capital Intensive Technology in a Capital-Poor Country. Cambridge, MA: MIT Press.

Findley, Ronald (1970) Trade and Specialization. Baltimore, MD: Penguin Books.

Finelli, Anton "Business Development, Venturing, and Risk: An Overview." CCED Newsletter, February-March 1978.

Fisher, Franklin M. (1972) The Economic Theory of Price Indices. New York: Academic Press.

_____ "Embodied Technical Change and the Existence of an Aggregate Capital Stock." Review of Economic Studies, Vol. 32, October 1965.

Fishlow, A. (1965) American Railroads and the Transformation of the Ante-Bellum Economy, Cambridge, MA: Harvard University Press.

_____ (1966) "Productivity and Technological Change in the Railroad Sector, 1840-1910." In Output, Employment, and Productivity in the United States After 1880. New York: Columbia University Press for National Bureau of Economic Research.

Fite, Emerson D. (1910) Social and Industrial Conditions in the North During the Civil War. New York: Macmillan Company.

Floor, W.N. (1977) "A Critical Review of Activities of the U.N. System on Appropriate Technology." Expert Meeting on International Action for Appropriate Technology.

Fogel, Rogert W. (1964) Railroads and American Economic Growth: Essays in Econometric History. Baltimore, MD: The Johns Hopkins Press.

_____ andEngerman, Stanley "A Model for the Explanation of Industrial Expansion during the Nineteenth Century: With an Application to the American Iron Industry." Journal of Political Economy, Vol. LXXVII, 3, May/June 1969.

Foley, Gerald (1976) The Energy Question. Baltimore, MD: Penguin Books.

Ford Foundation (1974) A Time to Choose. Cambridge, MA: Ballinger.

Fowles, Jib., ed. (1978) Handbook of Futures Research. Westport, CT: Greenwood Press.

Frank, Andre (1973) Capitalism and Underdevelopment in Latin America. New York: Monthly Review Press.

_____ (1970) Development of Underdevelopment. New York: Monthly Review Press.

Frankel, Francine R. (1971) Indian Green Revolution: Economic Gains and Political Costs. Delhi: Oxford University Press.

Frankel, M. "Obsolescence and Technological Change." American Economic Review, Vol. 45, June 1955.

French, David (1977) Appropriate Technology in Social Context: An Annotated Bibliography. Washington, D.C.: U.S. Agency for International Development.

Friedman, Milton (1962) Capitalism and Freedom. Chicago, IL: University of Chicago Press.

_____ "Foreign Economic Aid: Means and Objectives." Yale Review, June 1958, 500-16.

Frierei, Paulo (1973) Education for Critical Consciousness. New York: Seabury Press.

_____ (1970) Pedagogy of the Oppressed. New York: Seabury Press.

Fritsch, Albert J. ed. (1976) Solar Energy: One Way to Citizen Control. Washington, D.C.: Center for Science in the Public Interest.

Fromm, Erich (1973) The Anatomy of Human Destructiveness. New York: Holt, Rinehart & Winston, Inc.

_____ (1963) Marx's Concept of Man. New York: Frederick Ungar Publishing Co.

Galbraith, John K. (1974) Economic Development. Geneva: Houghton Mifflin Co.

_____ (1973) Economics and the Public Purpose. Boston: Houghton Mifflin Co.

_____ (1965) The New Industrial State. Boston: Houghton Mifflin Co.

Galenson, Walter "Economic Development and the Sectoral Expansion of Employment." International Labor Review, Vol. 87, No. 6, June 1963, 505-19.

_____ andLeibstein, Harvey "Investment Criteria, Productivity and Economic Development." Quarterly Journal of Economics, August 1955, 343-70.

Galtung, Johan (1978) Development, Environment and Technology: Towards a Technology for Self-Reliance. Geneva: United Nations Conference on Trade and Development.

_____ The True Worlds. New York: The Free Press, forthcoming.

Gandhi, M.K. (1969) All Men are Brothers. Paris: UNESCO.

_____ (1957) Economic and Industrial Life and Relations, 3 Vols., Ahmedabad: Navjivan Publishers House.

_____ (1967) Political and National Life and Affairs, 3 Vols., Ahmedabad: Navjivan Publishers House.

_____ (1952) Rebuilding Our Village. Ahmedabad: Navjivan Publishers House.

Gates, Paul W. (1960) The Farmer's Age: Agriculture 1815-1860: The Economic History of the United States, Vol. III. New York: Holt, Rinehart, and Winston.

Gemmill, Gordon and Eicher, Carl (1973) The Economics of Farm Mechanization and Processing in Developing Countries. New York: The Agricultural Development Council, Inc.

Genovese, Eugene D. (1965) The Political Economy of Slavery. New York: Random House.

Georgescu-Roegen, N. "Economic Theory and Agrarian Economics." Oxford Economic Papers, Vol. 12, February 1960, 1-40.

_____ (1971) The Entropy Law and the Economic Process. Cambridge, MA: Harvard University Press.

German Foundation for Developing Countries (1972) Development and Dissemination of Appropriate Technologies in Rural Areas, Berlin.

Germidis, Dimitri and Brochet, Christine (1975) Le Prix du Transfert de

Technologies dans les Pays en Voie de Development. Organization for Economic Cooperation and Development, Development Center, Etude Special, No. 5, Paris: Organization for Economic Cooperation and Development.

Gerschenkron, A. (1962) Economic Backwardness in Historical Perspective. Cambridge, MA: Harvard University Press.

Gill, T. Richard (1963) Economic Development: Past and Present. Englewood Cliffs, NJ: Prentice-Hall.

Gillete, R. "Latin America: Is Imported Technology Too Expensive?" Science, 181-4-44, July 1973.

Giral, Jose Barnes (1973) Appropriate Chemical Technologies for Developing Economies. Mexico City: Universidad Nacional Autonoma de Mexico.

Goldschmidt, Walter (1978) As You Sow: Three Studies in the Social Consequences of Agribusiness. Montclair, NJ: Allanheld, Osmun & Co.

Goodman, Percival (1977) The Double E. New York: Doubleday.

Gordon, R.A. (1961) Business Leadership in the Large Corporation. Berkeley, CA: University of California Press.

_____ (1974) Economic Instability and Growth: The American Record. New York: Harper & Row.

_____ andGordon, M.S. (1967) Toward a Manpower Policy. New York: Praeger.

Gotsch, Carl "Technical Change and the Distribution of Income in Rural Areas." American Journal of Agricultural Economics. Vol. 54, May 1972, 326-341.

Goulet, Denis (1971) The Cruel Choice: A New Concept in the Theory of Development. New York: Atheneum.

_____ (1978) Looking at Guinea-Bissau. Washington, D.C. O.C.D.

_____ (1974) A New Moral Order: Development Ethics and Liberation Theology. Maryknoll NY: Orbis Books.

_____ (1977) The Uncertain Promise: Value Conflict in Technology Transfer. New York: IDOC.

Government of India (1970) Fourth Five Year Plan. New Delhi.

_____ (1976) Fuel Policy Committee Report. New Delhi.

_____ Planning Commission (1972) Report on the Task Force on Integrated Rural Development Projects in Canal Irrigated Areas. New Delhi.

Greeley, Andrew M. (1977) No Bigger than Necessary: An Alternative to Socialism, Capitalism and Anarchism. New York: New American Library.

Green, H.A.J. "Embodied Progress, Investment, and Growth." American Economic Review, Vol. LVI, March 1966, 138-51.

Griffin, Keith (1974) The Political Economy of Agrarian Change: An Essay on the Green Revolution. Cambridge, MA: Harvard University Press.

Griliches, Zvi "Research Expenditures, Education and the Aggregate Agricultural Production Function." American Economic Review, Vol. 54, December 1964, 961-75.

Grossman, Richard and Danekar, Gail (1977) Jobs and Energy Environmentalists for Full Employment. Washington, D.C.

Gurley, John G. (1976) China's Economy and the Maoist Strategy. New York: Monthly Review Press.

_____ "Maoist Economic Development: The New Man in New China." The Center Magazine, May 1970, 25-33.

_____ andShaw, Edward S. "Financial Structures and Economic Development." Economic Development and Cultural Change, Vol. 15, April 1970, 257-68.

Haavelmo, T. (1960) A Study in the Theory of Investment. Chicago: University of Chicago Press.

Habakkuk, N.J. (1962) American and British Technology in the Nineteenth Century. Cambridge: Cambridge University Press.

_____ (1976) Poverty Curtain. New York: Columbia University Press.

Hag, Mahbubul (1971) "Employment in the 1970's: A New Perspective." International Development Review, 9-13.

Hagen, Everett E. (1968) The Economics of Development. Homewood, IL: Richard D. Irwin.

_____ (1962) On the Theory of Social Change. Homewood, IL: The Dorsey Press, Inc.

_____ "The Process of Economic Development." Economic Development and Cultural Change, Vol. 5, April 1957, 193-215.

Hahn, Albert V.G. (1974) Towards a Reappraisal of the Petrochemical Industry: Technology and Economics. Occasional Paper No. 2. Paris: Organization for Economic Cooperation and Development.

Hahn, Frank H. "On Two-Sector Growth Models." Review of Economic Studies. Vol. XXXII, October 1965, 339-45.

_____ (1971) Readings in the Theory of Growth. New York: St. Martins Press.

_____ andMatthews, R.C.O. "The Theory of Economic Growth: A Survey." EconomicJournal, Vol. LXXIV, December 1964, 779-902.

Hall, G.R. and Johnson, R.E. (1970) "Transfer of United States Aerospace Technology to Japan." Edited by J.R. Vernon, The Technology Factor in International Trade. New York: National Bureau of Economic Research.

Hall, R.E. "Technical Change and Capital from the Point of View of the Dual." Review of Economic Studies, Vol. XXXV, January 1968, 35-46.

Hammond, Anthony (1856) "Use of Reaping Machines." Journal of the Royal Agricultural Society of England, Vol. XVII.

Hannon, Bruce M. (1976) Energy and Labor Demand in the Conserver Society. Center for Advanced Computation. Urbana, IL: University of Illinois Press.

Harberler, Gottfried (1959) "International Trade and Economic Development." National Bank of Egypt Fiftieth Anniversary Commemoration Lectures.

Harbison, Frederick "Entrepreneurial Organization as a Factor in Economic Development." Quarterly Journal of Economics, Vol. 70, August 1956, 364-79.

Harcourt, G.C. "Some Cambridge Controversies in the Theory of Capital." Journal of Economic Literature, Vol. VII, June 1969, 369-405.

Harman, Willis W. "Chronic Unemployment: An Emerging Problem of Postindustrial Society." The Futurist, August 1978, 209-14.

_____ "The Coming Transformation" The Futurist, April 1977.
Harral, Clell G., et al. Study of the Substitution of Labor and Equipment in Civil Construction: Phase II Final Report, 3 Vols. Washington, D.C. International Bank for Reconstruction and Development, 1974.
Harris, J. and Todaro, M. "Migration, Unemployment and Development." American Economic Review, 60, March 1970, 126-42.
Harrod, R.F. (1969) Money. London: Macmillan.
_____ (1948) Towards a Dynamic Economics. London: Macmillan.
_____ The Trade Cycle. Oxford: The Clarendon Press, 1936.
Hauser, Philip M., ed. (1969) The Population Dilemma, 2nd Edition, Englewood Cliffs, NJ: Prentice-Hall.
Hayek, F.A. (1975) The Pure Theory of Capital. Chicago, IL: University of Chicago Press.
Hayes, Denis (1977) Rays of Hope: The Transition to a Post-Petroleum World. New York: W.W. Norton.
_____ (1978) Repairs, Reuse, Recycling-First Steps Toward a Sustainable Society. Washington, D.C,: Worldwatch Institute.
Hayter, Teresa (1971) Aid as Imperialism. Middlesex, England: Penguin Books.
_____ (1966) French Aid.London: Overseas Development Institute, Ltd.
Hazari,Bharat K. and Krishnamurty, J. "Employment Implications of India's Industrialization: Analysis in an Input-Output Framework." Review of Economics and Statistics, Vol. LII, May 1970, 181-6.
Heilbroner, R.L. (1976) Business Civilization in Decline. New York: W.W. Norton.
_____ (1963) The Great Ascent. New York: Harper & Row.
_____ (1974) An Inquiry into the Human Prospect. New York: W.W. Norton.
Helleiner, G.K. "The Role of Multinational Corporations in the Less Developed Countries." Trade in Technology, World Development, Vol. 3, April 1975.
Heller, R.H. (1973) International Trade Theory and Empirical Evidence, 2nd Edition, Chapel Hill, NC: Preston-Hill.
Henderson, Hazel (1978) Creating Alternative Futures: The End of Economics. New York: Berkley.
Hesburg, Theodore M. (1974) The Humane Imperative: A Challenge for the Year 2000. New Haven, CT: Yale University Press.
Hibbs, Douglas A. (1973) Mass Political Violence: A Cross-National Causal Analysis. New York: John Wiley & Sons.
Hickman, B.G. (1965) Investment Demand and U.S. Economic Growth. Washington, D.C.: Brookings Institution.
Hicks, Sir John R. (1965) Capital and Growth. Oxford: The Clarendon Press.
_____ (1932) The Theory of Wages. London: Macmillan.
_____ (1975) Value and Capital: Inquiry into Some Fundamental Principles of Economic Theory. Oxford: Oxford University Press.
Hicks, W. Whitney "A Note on the Burden of Dependency in Low Income Areas." Economic Development and Cultural Change. Vol. 13, January 1965, 233-5.
Higgins, Benjamin (1968) Economic Development, Rev. Ed. New York: W.W. Norton & Co., Inc.

Hirsh, Fred (1977) The Social Limits to Growth. Cambridge, MA: Harvard University Press.

_____ Michael W. Doyle; and Morse, Edward L. (1978) Alternatives to Monetary Disorders. New York: McGraw-Hill.

Hirshman, Albert O. "The Political Economy of Import-Substituting Industrialization in Latin America." Quarterly Journal of Economics, Vol. 83, February 1968, 1-32.

_____ (1958) The Strategy of Economic Development. New Haven, CT: Yale University Press.

Hollins, Elizabeth Jay (1966) Peace is Possible. New York: Grossman.

Hoos, Ida R. "Systems Analysis as Technology Transfer." Journal of Dynamic Systems, Measurement, and Control. 96, March 1974, 1-5.

Horie, Y. (1965) "Modern Entrepreneurship in Meiji Japan." The State and Economic Enterprise in Japan. Edited by W.W. Lockwood. Princeton, NJ: Princeton University Press.

Hoselitz, Bert F., ed. (1960) Theories of Economic Growth. New York: The Free Press.

Host-Madsen, Paul "Balance of Payments Problems of Developing Countries." Finance and Development, Vol. 4, No. 2, June 1967, 118-24.

Houthakkar, Hendrick S. and Taylor, Lester D. (1970) Consumer Demand in the United States. Cambridge: Harvard University Press.

Hoyt, Homer (1933) One Hundred Years of Land Values in Chicago. Chicago, IL: Chicago University Press.

Hunter, John M. "Underdeveloped Nations." Business Topics, Vol. 10, No. 2, Spring 1962, 17-30.

Huntington, Samuel P. (1968) Political Order in Changing Societies. New Haven, CT: Yale University Press.

Huq, Mahbub-ul (1976) The Poverty Curtain. New York: Columbia University Press.

Illich, Ivan (1969) Celebration of Awareness. New York: Doubleday.

_____ (1970) Deschooling Society. New York: Harper & Row.

_____ (1974a) Energy and Equity. Harper & Row or Perenenial Library Edition.

_____ (1974b) Medical Nemesis; The Expropriation of Health. New York: Random House.

_____ (1973) Tools for Conviviality. New York: Harper & Row.

Inkeles, Alex and Smith, David H. (1974) Becoming Modern: Individual Change in Six Developing Countries. Cambridge, MA: Harvard University Press.

International Institute for Environment and Development (1978) Banking on the Biosphere? The Environmental Policies and Practices of Nine Multilateral Aid Institutions. Washington, D.C. IIED.

International Labor Office (1977) Employment, Growth and Basic Needs: A One-World Problem. New York: Praeger Publishers.

_____ (1978) Technology, Employment and Basic Needs. Geneva: International Labor Organization, 1978. (ILO Overview Paper prepared for the United Nations Conference on Science and Technology for Development.)

_____ (1978) Report Prepared for the Second Tripartite Technical

Meeting for the Food Products and Drink Industries. Geneva: ILO.

Islam, Nurul, ed. (1974) Agricultural Policy in Developing Countries. New York: Halsted Press.

Jack, Homer, ed. (1968) World Religions and World Peace. Boston: Beacon.

Jackson, Barbara Ward (1962) The Rich Nations and the Poor Nations. New York: W.W. Norton.

Jarrett, Henry, ed. (1966) Environmental Quality in a Growing Economy. Baltimore, MD: Johns Hopkins Press.

Jaszi, G. "An Improved Way of Measuring Quality Change." Review of Economics and Statistics, Vol. XLIV, August 1962, 332-5.

Jedlicka, Allen D. (1977) Organization for Rural Development: Risk Taking and Appropriate Technology. New York: Praeger Publishers, Inc.

Jenkins, G. (1975) Non-Agricultural Choice of Technique: An Annotated Bibliography of Empirical Studies. Oxford: Institute of Commonwealth Studies.

Jequier, Nicolas, ed. (1976) Appropriate Technology, Problems and Promises. Paris: Organization for Economic Cooperation and Development.

Johnson, H.G. (1967) Economic Policies Towards Less Developed Countries. New York: Frederick A. Praeger.

Johnson, Jack, ed. (1976) The Application of Technology in Developing Countries. Office of Arid Lands Studies. Tucson: University of Arizona Press.

Jones, E.L. "The Agricultural Labor Market in England, 1793-1872." Economic History Review, 2nd Series, Vol. 17, 1964.

_____ "The Development of a Dual Economy." Economic Journal, 1961, 309-34.

Jorgenson, D.W. (1967b) "Surplus Agricultural Labor and the Development of a Dual Economy." Oxford Economic Papers, Vol. 19, November 1967, 281-312.

_____ (1967a) "The Theory of Investment Behavior." Edited by R. Faber, Determinants of Investment Behavior. New York: Columbia University Press for NBER.

_____ andChristensen, L.R. "U.S. Real Product and Real Factor Input, 1929-1967." Review of Income and Wealth, Vol. XVI, I, March 1970.

_____ andGriliches, Z. "The Explanation of Productivity Change." Review of Economic Studies, Vol. XXXIV, July 1967.

_____ "Sources of Measured Productivity Change: Capital Input." American Review, Papers and Proceedings, Vol. LXI, May 1966, 50-61.

Kahn, Herman and Martel, Leon (1976) The Next 200 Years. New York: William Morrow.

Kaldor, N. (1967) Strategic Factors in Economic Development. Ithaca: Cornell University Press.

Kamenetzky, Mario (1978) "Psycho-Social Stability and the Appropriateness of Technology." Paper presented to the Symposium Towards Integration of Science and Technology with Development Needs of the LDCs. Florida International University, April.

Kelley, Alan C.; Williamson, Jeffrey; and Chatham, Russell J. (1972) Dualistics Economic Development. Chicago: University of Chicago Press.

Kendrick, J.W. (1961) Productivity Trends in the United States. Princeton, NJ: Princeton University Press for NBER.

Kennedy, Charles (1962b) "The Character of Improvements and of Technical Progress." Economic Journal, LXXII, December 1962, 899-911.

_____ (1962a) "Harrod on 'Neutrality.' " Economic Journal, LXXII, March 1962, 249-50.

_____ "Induced Bias in Innovation and the Theory of Distribution." Economic Journal, Vol. LXXIV, September 1964.

_____ "Technical Progress and Investment." Economic Journal, LXXI, June 1961, 292-9.

_____ "The Valuation of Net Investment." Oxford Economic Papers, VII, I, February 1955, 35-46.

Keohane, Robert O. and Nye, Joseph S. (1977) Power and Interdependence: World Politics in Transition. Boston, MA: Little, Brown & Co.

Keynes, J.M. (1936) The General Theory of Employment Interest and Money. London: Macmillan.

Khan, M.Y. "Employment Growth in Organized Sector, 1961-73." Reserve Bank of India Bulletin, February 1975, 129-38.

Kilby, Peter "Farm and Factory: A Comparison of the Skill Requirements for the Transfer of Technology." The Journal of Development Studies, 9, October 1972, 63-70.

Kim, Young Chim "Sectoral Output-Capital Ratios and the Level of Economic Development: A Cross-Sectional Comparison of Manufacturing Industry." Review of Economics and Statistics, Vol. 51, No. 4, November 1969, 453-58.

Kindleberger, Charles P. (1965) Economic Development. 2nd ed., New York: McGraw-Hill.

King, Timothy (1974) Population Policies and Economic Development: A World Bank Staff Report. Baltimore, MD: Published for the World Bank by the Johns Hopkins Press.

Klein, L.R.; Ball, R.J.; Hazelwood, A.; and Vandome, P. (1961) An Econometric Model of the United Kingdom. Oxford: Basil Blackwell.

Knoppers, Antonie T. "Development and Transfer of Marketable Technology in the International Corporation: A New Situation." Applied Science and World Economy. Paper presented to the Committee on Science and Astronautics, U.S. House of Representatives, for the Ninth Meeting of the Panel on Science and Technology, February 1968, Washington, D.C.: U.S. Government Printing Office, 21-34.

Korea Institute of Science and Technology (1973) Final Report of a Study of the Scope for Capital-Labor Substitution in the Mechanical Engineering Sector. Seoul: Korea Institute of Science and Technology.

Kothari, Rajni (1975) Footsteps into the Future. New York: The Free Press.

Krause, Walter (1961) Economic Development. Belmont, CA: Wadsworth Publishing Co.
Krishna, Raj "Unemployment in India." Economic and Political Weekly, March 3, 1973.
Krishnamurty, J. (1974) Indirect Employment Effects of Investment in Industry. World Employment Programme Research, WEP-22/WP6. Geneva: International Labor Office.
Kunkel, John H. (1970) Society and Economic Growth. Fairlawn: Oxford University Press.
Kuznets, Simon (1953) Economic Change. New York: W.W. Norton.
_____ (1965) Economic Growth and Structure. New York: W.W. Norton.
_____ (1964) Postwar Economic Growth: Four Lectures. Cambridge, MA: Harvard University Press.
_____ (1959) Six Lectures on Economic Growth. New York: Free Press.
Lab, D. (1977) Men or Machines: A Philippines Case Study of Labor-Capital Substitution in Road Construction. Geneva: ILO.
Lachmann, Karl E. "The Role of International Business in the Transfer of Technology to Developing Countries," Paper presented to the American Society of International Law, Sixtieth Annual Meeting, Washington, D.C., April 28, 1966.
Lakey, George (1973) Strategy for a Living Revolution. New York: Freeman & Co.
Lancaster, K. "A New Approach to Consumer Theory." Journal of Political Economy, LXXIV, April 1966, 132-56.
Landes, David S. "Factor Costs and Demand, Determinants of Economic Growth." Business History, Vol. VII, 1, January 1965.
_____ (1969) The Unbound Prometheus. Cambridge: Cambridge University Press.
Langdon,S. "Multinational Corporations' Taste Transfer and Underdevelopment: A Case Study from Kenya." Review of African Political Economy, January-April 1975.
Lappe, Francis Moore; and Collins, Joseph (1979) Food First: Beyond the Myth of Scarcity. Boston: Houghton Mifflin.
Leaner, Daniel and Schram, Wilbur, eds. (1967) Communication and Change in the Developing Countries. Honolulu: East-West Center Press.
Lebergott, Stanley (1964) Manpower and Economic Growth: The United States Record Since 1800. New York: McGraw-Hill Book Co.
Leff, Nathaniel (1968b) The Brazilian Capital Goods Industry, 1929-1964. Cambridge, MA: Harvard University Press.
_____ "Dependency Rates and Savings Rates." American Economic Review, Vol. 59, No. 5, December 1969, 886-96.
_____ (1968a) Economic Policy Making and Development in Brazil. New York: John Wiley & Sons.
Leibenstein, Harvey (1966b) "Allocative Efficiency vs. X-efficiency." American Economic Review, 56, June 1966, 392-415.
_____ (1976) Beyond Economic Man. Cambridge, MA: Harvard University Press.
_____ (1957) Economic Backwardness and Economic Growth. New York: John Wiley & Sons, Inc.

_____ (1966a) "Incremental Capital-Output Ratios and Growth Rates in the Short-Run." Review of Economics and Statistics, Vol. 48, February 1966, 20-7.
Leijonhufvud, A. (1968) On Keynesian Economics and the Economics of Keynes. New York: Oxford University Press.
Leontief, Wassily, ed. (1953) Studies in the Structure of the American Economy. New York: Oxford University Press.
Lerner, A.P. (1944) The Economics of Control. New York: Macmillan.
Letwin, William "Four Fallacies About Economic Development." Daedalus, Vol. 92, Summer 1963, 396-414.
Lewis, John P. (1962) Quiet Crisis in India. Washington, D.C.: The Brookings Institution.
Lewis, W.A. (1966) Development Planning: The Essentials of Economic Policy. New York: Harper & Row.
_____ "Economic Development with Unlimited Supplies of Labor." Manchester School, Vol. 22, May 1954, 139-91.
_____ "Unlimited Supplies of Labor: Further Notes." Manchester School, Vol. 26, January 1958, 1-32.
Linder, Staffan B. (1970) The Harried Leisure Class. New York: Columbia University Press.
Little, Ian, M.D. (1957) A Critique of Welfare Economics. Oxford: Clarendon Press.
_____ et al. (1970) Industry and Trade in Some Developing Countries. Oxford: Oxford University Press.
_____ andClifford, J.M. (1966) International Aid. Chicago: Aldine Publishing Co.
_____ andMirrlees, James (1969) Manual of Industrial Projects Analysis in Developing Countries, Social Cost-Benefit Analysis, Vol. II. Paris: Organization for Economic Cooperation and Development, Development Center.
Livingston, Dennis "Little Science Policy." Policy Studies Journal, Winter 1977.
_____ "Global Equilibrium and the Decentralized Community." Ekistics, September 1976, 173-176.
Lohia, Ram Maushar (1963) Marx, Gandhi and Socialism. Hyderabad: NAVA Hind Publications.
Long, W. Harwood "Development of Mechanization in English Farming." Agricultural History Review, Vol. II, 1963.
Lotz, Joergen R. and Morss, Elliot R. "A Theory of Tax Level Determinants for Developing Countries." Economic Development and Cultural Change, Vol. 18, No. 3, April 1970, 328-41.
Lovins, Amory B. (1977) Soft Energy Paths: Toward a Durable Place. Boston, MA: Ballinger Press.
Lundberg, Erik (1968) Instability and Economic Growth. New Haven, CT: Yale University Press.
_____ (1961) Produktivitet och Rantabilitet. Stockholm: P.A. Norstedt and Soner.
Lyaschenko, P.I. (1949) History of the National Economy of Russia to the 1917 Revolution. New York: Macmillan.
Machava, K.B., ed. (1969) International Development. Dobbs Ferry, NY: Oceana Press.

Machlup, Fritz (1967) The Production and Distribution of Knowledge in the United States. Princeton, NJ: Princeton University Press.

Macpherson, George and Jackson, Dudley "Village Technology for Rural Development: Agricultural Innovation in Tanzinia." International Labor Review, 3, February 1975, 105-26.

Maddison, Angus (1970) Economic Progress and Policy in Developing Countries. London: George Allen and Unwin.

Malcolm X (1965) Autobiography of Malcolm X. New York: Grove Press, Inc.

Management Analysis Center (1976) The Inducement of U.S. Firms to Adopt Products and Processes to Meet Conditions in Less Developed Countries. (Robert Stobaugh) Washington, D.C.: Agency for International Development.

Mandel, Ernest (1968) Marxist Economic Theory, 2 Vols. New York: Monthly Review Press.

Manser, W.A.P. (1974) "The Financial Role of Multinational Enterprises." In Multinational Enterprises. Edited by J.S.G. Wilson and C.F. Scheffer. Leiden, TheNetherlands: A.W. Sijhoff.

Marien, Michael (1977) "The Two-Visions of Post-Industrial Society." Paper presented to the American Sociological Association, Annual Meeting.

Markley, O.W. (1974) Changing Images of Man. Menlo Park, CA: Center for the Study of Social Policy, SRI International.

Marsden, K. (1971) "Progressive Technologies for Developing Countries." In Essays on Employment. Edited by W. Galenson, ILO.

Martin, Edwin J. (1972) Development Assistance: Efforts and Policies of the Members of the Development Assistance Committee. Organization for Economic Cooperation and Development Review Series, 1971. Paris: Organization for Economic Cooperation and Development.

Marx, K. (1918) A Contribution to the Critique of Political Economy. Trans. N.I. Stone. Chicago: Charles H. Kerr.

_____ Gundrisse. Penguin Books, Middlesex, England (1973).

Mason, Edward S. and Asher, Robert E. (1973) The World Bank Since Bretton Woods. Washington, D.C.: The Brookings Institution.

Mason, R. Hal "The Multinational Firm and the Cost of Technology to Developing Countries." California Management Review, 15, 5-13, Summer 1973.

_____ (1970) The Transfer of Technology and the Factor Proportions Problem. The Philippines and Mexico, UNITAR Research Project No. 10.

Maslow, Abraham H. (1973) The Farther Reaches of Human Nature. New York: Viking Press.

_____ (1970) Motivation and Personality, Rev. Ed. New York: Harper & Row.

_____ (1962) Toward a Psychology of Being. Princeton, NJ: Van Nostrand.

Matthews, William H., ed. (1976) Outer Limits and Human Needs: Resource and Environmental Issues of Development Strategies. Uppsala: Dag Hammarskjold Foundation.

McCallum, Bruce (1977) Environmentally Appropriate Technology: Renewable Energy and Technologies for a Conserver Society in Canada. Ottawa: Department of Fisheries and the Environment, Government of Canada.

McClelland, David C. (1967) The Achieving Society. New York: Free Press.

McCloskey, D.N., ed. (1971) Essays on a Mature Economy: Great Britain After 1840. London: Methuen and Co., Ltd.

McGouldrick, Paul F. (1968) New England Textiles in the Nineteenth Century, Profits and Investment. Cambridge, MA: Harvard University Press.

McHale, John and Magda, C. (1978) Basic Human Needs: A Framework for Action. New Brunswick, NJ: Transaction Books.

McKinnon, R.I. "Foreign Exchange Constraints in Economic Development and Efficient Aid Allocation." Economic Journal, 1964, 399-409.

McPolue, G. and Carr, M. (1975) "Mass Production by the Masses." In Technology: A Critical Choice for Developing Countries. Intermediate Technology Group.

Mead, Margaret (1972) Twentieth Century Faith, Hope and Survival. New York: Harper & Row.

_____ (1975) World Enough: Rethinking the Future. Boston: Little, Brown, and Co.

Meade, J.E. (1968a) The Growing Economy, Vol. II. Principles of Political Economy. Albany: State University of New York Press.

_____ (1968b) A Neo-Classical Theory of Economic Growth, 2nd Ed. London: Allen and Unwin.

_____ "The Rate of Profit in a Growing Economy." Economic Journal LXXIII, December 1963, 655-74.

Meadows, Donella; Meadows, Dennis L.; Randers, Jorgen; and Behrens, William W., III (1972) The Limits to Growth. Washington, D.C.: Potomac Associates.

Meir, Gerald M., ed. (1970) Leading Issues in Development Economies, 2nd Ed. Fairlawn: Oxford University Press.

Melman, Seymour (1965) Our Depleted Society. New York: Holt, Rinehart & Winston, Inc.

Memmi, Albert (1967) Colonizer and the Colonized. Boston: Beacon Press.

_____ (1968) Dominated Man Boston: Beacon Press.

Mendis, D.L.O. (1975) Planning the Industrial Revolution in Sri Lanka. Organization for Economic Cooperation and Development, Development Center, Occasional Paper No. 4. Paris: Organization for Economic Cooperation and Development.

Mendlovitz, Saul H., ed. (1975) On the Creation of a Just World Order. New York: The Free Press.

Merhan, M. (1967) Technological Dependence, Monopoly, and Growth. Elmsford, NY: Pergamon Press.

Merrill, Richard, ed. (1976) Radical Agriculture. New York: Harper & Row.

Mesarovic, Mihajlo and Pestel, Edward (1974) Mankind at the Turning Point. New York: E.P. Dutton & Co.

Meursinge, J. "Practical Experience in the Transfer of Technology." Technology and Culture, 12, July 1971, 469-70.

Michaelis, Michael "The Environment for Technology Transfer." Proceedings of a Conference on Technology Transfer and Innovation, sponsored by the National Planning Association and National Science Foundation, May 15-17, 1966, Washington, D.C.: U.S. Government Printing Office, 1967, 76-83.

Michie, Barry H. "Variations in Economic Behavior and the Green Revolution: An Anthropological Perspective." Economic and Political Weekly, VI II, June 30, 1973, India: Sameeksah Trust Publications, A-67-A75.

Miles, Rufus E., Jr. (1976) Awakening from the American Dream: The Social and Political Limits to Growth. New York: Universe Books.

Miller, Morris "The Scope and Content of Resources Policy in Relation to Economic Development." Land Economics, Vol. 37, No. 4, November 1961, 291-310.

Millikan, Max F. and Hapgood, David (1967) No Easy Harvest. Boston, MA: Little, Brown & Co.

Minhas, B.S. (1974) Planning and the Poor. New Delhi: S. Chand and Co.

Mische, Gerald and Mische, Patricia (1977) Toward a Human World Order. New York: Paulist Press.

Mishan, E.J. (1970) Technology and Growth: The Price We Pay. New York: Praeger Publishers, Inc.

Mishikawa, Shunsaku "Unemployment Statistics Compilation Needs Revision." The Japan Economic Journal/Nihon Keizai Shimbun. Tokyo, No. 13, May 6, 1975, 3.

Montgomery, John D. (1974b) Allocation of Authority in Land Reform Programs: A Comparative Study of Administrative Processes and Outputs. Research and Training Network Reprint Series, March 1974, New York: Agricultural Development Council, Inc.

_____ (1974a) Technology and Civic Life: Making and Implementing Development Decisions. Cambridge, MA: MIT Press.

Morawetz, D. "Elasticities of Substitution in Industry." World Development, Vol. 4, No. 7, January 1976.

_____ "Employment Implications of Industrialization in Developing Countries: A Survey." Economic Journal, Vol. 84, September 1974.

_____ (1975) "Import Substitution, Employment and Foreign Exchange in Columbia: No Cheers for Petrochemicals." In The Choice of Technology in Developing Countries: Some Cautionary Tales, by C. Peter Trimmer, John Woodward Thomas, Louis T. Wells, Jr., and David Morawetz, 95-105. Cambridge, MA: Harvard University, Center for International Affairs.

Morehouse, Ward (1968) Science and the Human Condition in India and Pakistan. New York: Rockefeller University Press.

_____ (1977) Science Technology and the Global Equity Crisis: New Direction for U.S. Policy. Stanley Foundation Strategy for Peace Conference.

_____ andSigurdson, John "Science, Technology and Poverty." Bulletin of the Atomic Scientists. December 1977.

Morgan, T.; Betz, G.W.; and Choudry, N.K., eds. (1963) Readings in Economic Development. Belmont, CA: Wadsworth, 1963.

Morris, David "Appropriate Technology and Community Economic Development." CCED Newsletter, April-May 1977, 1-9.

_____ andHess, Karl (1975) Neighborhood Power: The New Localism. Boston, MA: Beacon Press.

Morrison, E. Denton (1978) "Energy and Interdependence," Paper prepared for the Woodrow Wilson Center, Smithsonian Institution, Washington, D.C.

Mountjoy, Alan B. (1967) Industrialization and Underdeveloped Countries, Rev. Ed. Chicago, Ill: Aldine Publishing Co.

Mowlana, Hamid (1975) "The Multinational Corporation and the Diffusion of Technology." In The New Sovereigns, by Abdul A. Said and L.R. Simmons. Englewood Cliffs, NJ: Prentice Hall.

Mumford, Lewis (1967) The Myth of the Machine: Techniques and Human Development. New York: Harcourt Brace Jovanovich, Inc.

Mureithe, L.P. "Demographic and Technological Variables in Kenya's Employment Scene." Eastern Africa Economic Review, Vol. 6, June 1974.

Myint, Hla (1971) Economic Theory and Underdeveloped Countries. Fairlawn: Oxford University Press.

_____ (1964) The Economics of Developing Countries. New York: Praeger.

Myrdal, Gunnar (1974) Against the Stream: Critical Essays on Economics. New York: Random House.

_____ (1968) Asian Drama: An Inquiry into the Poverty of Nations. 3 Vols., New York: Pantheon Books.

_____ (1970) Challenge of World Poverty: A World Poverty Program in Outline. New York: Pantheon Books.

_____ (1978) An International Economy: Problems and Perspectives. Westport, CT: Greenwood Press.

_____ (1957) Rich Lands and Poor. New York: Harper & Row.

Nance, John (1975) The Gentle Tasaday. New York: Harcourt Brace Jovanovich, Inc.

Nasbeth, L. and Ray, G.F. (1974) The Diffusion of New Industrial Processes. Cambridge, England: Cambridge University Press.

Nash, Hugh, ed. (1978) Progress as if Survival Mattered: A Handbook for a Conserver Society. San Francisco, CA: Friends of the Earth.

Nash, Manning "Some Notes on Village Industrialization in South and East Asia." Economic Development and Cultural Change. III, No. 3, April 1955, 271-7.

National Bureau of Economic Research (1962) The Rate and Direction of Inventive Activity. Princeton, NJ: Princeton University Press.

National Council of Applied Economic Research (1971) Foreign Technology and Investment: A Study of Their Role in India's Industrialization. New Delhi.

Nayudamma, Y. "Promoting the Industrial Application of Research in an Underdeveloped Country." Minerva, No. 4, Spring 1967, 323-39.

Negandhi, Anana R. and Prasad, S. Benjamin (1975) The Frightening Angels: A Study of U.S. Multinationals in Developing Nations. Kent, OH: The Kent State University Press.

Nell, Edward J. "Theories of Growth and Theories of Value." Economic

Development and Cultural Change. Vol. 16, October 1967, 15-26.

Nelson, Richard R. "Less Developed Countries – Technology Transfer and Adaptation: The Role of the Indigenous Science Community." Economic Development and Cultural Change. 23, October 1974, 61-77.

_____ et al. (1967) Technology, Economic Growth and Public Policy. Washington, D.C.: The Brookings Institution.

_____ andWinter, Sidney G. "Towards an Evolutionary Theory of Economic Capabilities." American Economic Review, Vol. LXIII, 2 May 1973.

Nerfin, Marc, ed. (1977) Another Development: Approaches and Strategies. Uppsala, Sweden: Hammarskjold Foundation.

Nerlove, Marc (1967) The Theory and Empirical Analysis of Production. Edited by M. Brown. XXXI, New York: National Bureau of Economic Research Studies in Income and Wealth.

Nerrill, Richard, ed. (1976) Radical Agriculture. New York: Harper & Row.

_____ andGage, Thomas, eds. (1978) Energy Primer: Solar, Water, Wind and Biofuels. New York: Dell Delta.

Nicholson, J.L. "The Measurement of Quality Changes." Economic Journal, LXXVII, September 1967, 512-30.

Norman, Colin (1978) Soft Technologies, Hard Choices. Washington, D.C.: Worldwatch Institute.

North, D.C. (1961) The Economic Growth of the United States, 1790-1860. Englewood Cliffs, NJ: Prentice-Hall.

North, Michael, ed. (1977) Time Running Out: The Best of "Resurgence." New York: Universe Books.

Novack, David E. (1974) Transfer of Technology for Small Industries. Paris: Organization for Economic Cooperation and Development.

_____ andLekachman, Robert, eds. (1964) Development and Society. New York: St. Martin's Press.

Nurkse, Ragnar (1953) Problems of Capital Formation in Underdeveloped Countries, Oxford, England: Blackwell.

Nutter, G. Warren "On Measuring Economic Growth." Journal of Political Economy, Vol. 65, February 1957, 51-63.

Nyerere, Julius (1974) Man and Development, Oxford University Press.

Ohlin, Goran (1967) Population Control and Economic Development. Development Center of the Organization for Economic Cooperation and Development.

Ohun, Bernard and Richardson, Richard W., eds. (1961) Studies in Economic Development. New York: Holt Rinehart & Winston.

Okabe, Naoaki "Present Unemployment Can Become Very Serious." The Japan Economic Journal/Nihon Keizai Shimbun, Tokyo, No. 13, February 25, 1975, 1.

Ophuls, William (1977) Ecology and the Policies of Scarcity. San Francisco, CA: Freeman.

Organization for Economic Cooperation and Development (1974) Choice and Adaptation of Technology in Developing Countries: An Overview of Major Policy Issues. Paris.

_____ Development Center (1974-75) Low-Cost Technology: An Inquiry

Into Outstanding Policy Issues: Interim Report of the Study Sessions Held in Paris, 17-20, September 1974, Paris.

Oshima, Harry T. "Economic Growth and the 'Critical Minimum Effort.'" Economic Development and Cultural Change, Vol. 7, July 1959, 467-476.

_____ "Labor Force Explosion and the Labor Intensive Sector in Asian Growth." Economic Development and Cultural Change, 19, January 1971, 161-83.

Packard, Vance (1972) A Nation of Strangers. New York: David McKay Co., Inc.

Paglin, Morton "Surplus Agricultural Labor and Development: Facts and Theories." American Economic Review, Vol. 55, September 1965, 815-33.

Papanek, Gustav F. (1967) Pakistan's Development: Social Goals and Private Incentives. Cambridge: Harvard University Press.

Parker, William N. "Review of American and British Technology." The Business history Review, Vol. XXXVII, Spring/Summer 1963.

Pasinetti, L.L. "On Concepts and Measures of Changes in Productivity." Review of Economics and Statistics, Vol. XLI, August 1959, 270-82.

Patinkin, D. (1965) Money, Interest and Prices, 2nd Ed. New York: Harper & Row.

Pearson, Lester B. (1970) The Crisis of Development. New York: Praeger Publishers, Inc.

_____ (1969) Patterns in Development, Commission on International Development. New York: Praeger Publishers, Inc., 1969.

Perelman, Michael (1977) The Myth of Agricultural Efficiency. New York: Universe Books.

Phelps, Edmund S., ed. (1962) The Goal of Economic Growth. New York: W.W. Norton.

_____ andDrandakis, E.M. "A Model of Induced Invention, Growth, and Distribution." Economic Journal, Vol. LXXVI, December 1966.

Phillips, Walter "Technological Levels and Labor Resistance to Change in the Course of Industrialization." Economic Development and Cultural Change, Vol. XI, No. 3, April 1963.

Piasetzki, Peter. Energy and Environmentally Appropriate Technologies: A Selectively Annotated Bibliography. Council of Planning Librarians.

Pickett, J., ed. (1978) The Choice of Technology in Developing Countries. Elmsford, NY: Pergamon Press.

Pigou, A.C. (1935) The Economics of Stationary States. London: The Macmillan Co.

Pirages, Dennis Clard, ed. (1977) The Sustainable Society: Implications of Limited Growth. New York: Praeger Publishers, Inc.

Poblete, Juan Antonio and Harboe, Ricardo (1972) A Case on Transfer of Knowledge, in Water Resources Systems Planning, From a Developed Region to a Developing One and From Research to Application. Santiago, Chile: Universidad de Chile.

Posner, M.W. "International Trade and Technical Change." Oxford Economic Papers, Vol. 13, October 1961.

Prasad, K. (1963) Technological Choice Under Developmental Planning. New York: International Publications Service.

Prebisch, R. (1964) Towards a New Trade Policy for Development. New York: United Nations.

Prest, A.R. and Turvey, R. (1967) "Cost-Benefit Analysis: A Survey." Surveys of Economic Theory, Vol. III, London: The Macmillan Co.

Pronk, Jan, ed. (1975) Symposium on a New International Economic Order. The Hague: Ministry of Foreign Affairs.

Rain (1977) editors, Rainbook: Resources for Appropriate Technology. New York: Schocken Books.

Ramos, Joseph (1973) An Heterodoxical Interpretation of the Employment Problems in Latin America. Santiago, Chile: Programa Regional del Empleo Para America Latina y el Caribe.

Randhawa, M.S. (1974) Green Revolution. Delhi: Vikas Publishing House.

Ranis, G. (1976) Appropriate Technology in the Dual Economy: Reflections on Philippine and Taiwanese Experience. International Labor Office.

_____ ed. (1972) The Gap Between Rich and Poor Nations. London: Macmillan Press Ltd.

_____ (1972) Technology, Employment and Growth: The Japanese Experiment in Automation in Developing Countries. International Labor Office.

Rapping, Leonard "Learning and World War II Production Functions." Review of Economics and Statistics, Vol. 47, 1, 1965.

Rath, N. and Dandekar, V.M. (1971) Poverty in India. Bombay: Indian School of Political Economy.

Read, L.M. "The Measure of Total Factor Productivity Appropriate to Wage-Price Guidelines." Canadian Journal of Economics, I, May 1968, 349-58.

Reddaway, W.B. (1962) The Development of Indian Economy. Homewood, IL: Richard D. Irwin, Inc.

Reddy, A.K.N. "Alternative Technology: A View Point from India." Social Studies of Science, 1975, 331-42.

_____ "Choice of Alternative Technologies: Vital Tasks in Science and Technology Planning." Economic and Political Weekly, June 23, 1973, 1109-14.

Research Policy Program (1978) Technology Transformation of Developing Countries. Lund: University of Lund.

Reynolds, Lloyd "The Content of Economic Development." American Economic Review, Vol. 59, May 1969, 401-8.

Ricardo, D. (1951) On the Principles of Political Economy and Taxation. Edited by P. Sraffa and M. Dobb, The World and Correspondence of David Ricardo, Vol. I. Cambridge: Cambridge University Press.

Richards, J. (1872) A Treatise on the Construction and Operation of Woodworking Machines, London.

Rivers, Patrick (1976) The Survivalists. New York: Universe Books.

Roberts, John "Engineering Consultancy, Industrialization and Development." The Journal of Development Studies, 9, October 1972, 39-62.

Robertson, James (1976) Power, Money and Sex: Towards a New Social Balance. London: Marion Boyars.

_____ The Sane Alternative: Signposts to a Self-Fulfilling Future.

London: J.H. Robertson, 7th St., Ann's Villas, London WII 4RU, 1978.
_____ andRobertson, Carolyn (1978) The Small Towns Book: Show Me
the Way to Go Home. New York: Doubleday & Co.
Robertson, Ross M. (1964) History of the American Economy, 2nd Ed.
New York: Harcourt, Brace, and World.
Robinson, Joan (1955) The Accumulation of Capital, 3rd ed. London:
Macmillan.
_____ (1949) "The Classification of Inventions." Review of Economic
Studies, V., 1937-38, 139-2, Readings in the Theory of Income
Distribution. New York: Balkiston.
_____ (1962) Essays in the Theory of Economic Growth. London:
Macmillan.
Robinson, Richard D. (1964) Industrialization in Development Countries.
Cambridge: Cambridge University Overseas Studies Committee.
Rodriguez, C.A. "Trade in Technical Knowledge and the National
Advantage." Journal of Political Economy, 83, 212-35, February
1975.
Rogin, Leo (1931) The Introduction of Farm Machinery in Its Relation to
the Productivity of Labor in the Agriculture of the United States
During the Nineteenth Century. Berkeley: University of California
Press.
Rosenberg, N. "Economic Development and the Transfer of Technology:
Some Historical Perspectives." Technology and Culture, 11, 550-75,
October 1970.
_____ "Technological Change in the Machine Tool Industry, 1840-1910."
Journal of Economic History, vol. XXIII, December 1963.
_____ (1972) Technology and American Economic Growth. New York:
Harper & Row.
Rosenstein-Rodan, Paul N. "International Aid for Underdeveloped
Countries." Review of Economics and Statistics, vol. 43, May 1961.
_____ "Problems of the Industrialization of Eastern and South-Eastern
Europe." Economic Journal, 53, June-September 1943, 202-211.
Rostow, W.W. (1960) The Stages of Economic Growth. Cambridge:
Cambridge University Press.
Roszek, Theodore (1978) Person/Planet: The Creative Disintegration of
Industrial Society. Garden City, NY: Doubleday & Co.
Rottenberg, Simon "Income and Leisure in an Underdeveloped Country."
Journal of Political Economy, vol. 60, April 1952, 95-101.
Rubenstein, Albert H. et al. (1974) Research Priorities of Technology
Transfer to Developing Countries. Evanston, IL: Northwestern
University.
Ruttan, Vernon W. (1973) Technical and Institutional Transfer in
Agricultural Development. New York: Agricultural Development
Council.
_____ andArndt, Thomas M. (1975) Resource Allocation and
Productivity in National and International Agricultural Research.
Research and Training Network Seminar Report Series, September
1975. New York: Agricultural Development Council, Inc.
_____ andHayami, Yujiro (1971) Agricultural Development, An
International Perspective. Baltimore, MD: The Johns Hopkins Press.

Sagasti, Francisco R. "Underdevelopment, Science and Technology: The Point of View of the Underdeveloped Countries, Discussion Paper." Science Studies, 3, 1973, 47-59.

Said, Abdul and Simmons, L.R. (1974) The New Sovereigns. Englewood Cliffs, NJ: Prentice Hall.

Salter, W.E.G. (1966) Productivity and Technical Change, 2nd ed. Cambridge: Cambridge University Press.

Samuelson, Paul A. (1976) Economics, 10th Ed. New York: McGraw-Hill Book Co.

_____ "A Theory of Induced Innovation Along Kennedy-Weizsacker Lines." Review of Economics and Statistics. Vol. LVII, November 1965, 343-56.

_____ (1953) Foundations of Economic Analysis. Cambridge, MA: Harvard University Press.

Sandberg, Lars G. (1968) "American Rings and English Mules: The Role of Economic Rationality." Quarterly Journal of Economics, Vol. LXXXII, 4, November 1968.

Satin, Mark (1978) New Age Politics: Healing Self and Society: The Emerging New Alternative to Marxism and Liberalism. West Vancouver, B.C.: Whitecap Books, 1978.

Sau, Ranjit K. (1974) "Some Aspects of Inter Sectoral Resource Flow." Economic and Political Weekly Annual No. 1974, 1277-1284.

_____ (1973) Indian Economic Growth: Constraints and Prospects. New Delhi: Orient Longman Ltd., 1973.

Saul, S.B., ed. (1970) Technological Change: The United States and Britain in the Nineteenth Century. London: Methuen and Co., Ltd.

Schacht, Wendy H. and Renfro, William, with Keith Bea (1977) Appropriate Technology: A Review. Washington, D.C.: Congressional Research Service, Library of Congress, 1977.

Schiavo-Campo, Salvatore and Singer, Hans W., eds. (1970) Perspectives of Economic Development. Geneva: Houghton Mifflin.

Schmookler, Jacob (1966) Invention and Economic Growth. Cambridge, MA: Harvard University Press.

Schultz, Theodore (1964) Transforming Traditional Agriculture. New Haven, CT: Yale University Press.

Schumacher, E.F. (1977) A Guide for the Perplexed. New York: Harper & Row.

_____ (1973) Small is Beautiful: Economics as If People Mattered. New York: Harper & Row.

_____ (1972) "The Work of the Intermediate Technology Development Group in Africa." International Raliour Review, 106, 75-92, July 1972.

_____ andGillingham, Peter N. (1978) Good Work. New York: Harper & Row.

Schumpeter, Joseph A. (1934) The Theory of Economic Development. Cambridge, MA: Harvard University Press.

Science for the People (1974) China: Science Walks on Two Legs. New York: Avon Books.

Scitovsky, Tibor (1959) "Growth: Balanced or Unbalanced." Allocation of Economic Resources, M. Abramovitz, et al. Stanford, CA: Stanford University Press.

_____ (1976b) The Joyless Economy. New York: Oxford University Press.

_____ (1976a) "Science and Technology for Human Development: The Ambitious Future and the Christian Hope." Anticipation, 19, 1976.

_____ "Two Concepts of External Economics." Journal of Political Economy, Vol. 62, April 1954, 143-51.

Seers, Dudley "The Meaning of Development." International Development Review, 1969, 2-6.

Sen, Amartya K. (1962) Choice of Techniques: An Aspect of the Theory of Planned Economic Development, 2nd Ed. Oxford, England: Basil Blackwell.

_____ (1976) "Economic Development: Objectives and Obstacles." Paper presented at the Research Conference on the Lesson's of China's Development Experience for the Developing Countries, January 31-February 2, 1976, San Juan, Puerto Rico.

_____ (1975) Employment, Technology and Development. Oxford: Clarendon Press.

_____ ( ) "The Lessons of China's Development Experience for Developing Countries." Mimeo.

_____ (1973b) On Economic Inequality. New York: W.W. Norton.

_____ (1973a) "Poverty, Inequality and Unemployment." Economic and Political Weekly Special Issue 1973, 1457-64.

Senior, N.W. (1938) An Outline of the Science of Political Economy, New York: A.M. Kelley.

Serionich, R.C. (1974) Foreign Technology and Control in the Argentinian Industry. Brighton: Sussex University.

Service, Elman R. (1962) Primitive Social Organization. New York: Random House.

Sethi, J.D. (1978) Gandhi Today. Durham: North Carolina University Press.

_____ (1975) India in Crisis. Delhi: S. Chand and Co.

Sewell, J.N., et al. (1977) The U.S. and World Development Agenda. New York: Praeger.

Shackle, G.L.S. (1965) A Scheme of Economic Theory, Cambridge: Cambridge University Press.

Shannon, Lyle W., ed. (1957) Underdeveloped Areas. New York: Harper & Row.

Sheshinski, Eytan "Tests of the Learning by Doing Hypothesis." Review of Economics and Statistics, Vol. 49, 4, November 1967.

Shetty, M.C. (1963) Small-Scale and Household Industries in a Developing Country: A Study of Their Rationale, Structure and Operative Conditions. New York: Asia Publishing House.

Shields, Geoffery "The Multinationals." World Issues, December 1977, January 1978 and April-May 1978.

Shourie, Arun K. "India: An Arrangement at Stake." Economic and Political Weekly, June 22, 1974.

Simon, Arthur (1975) Bread for the World. New York: Paulist Press.

Simon, Paul and Simon, Arthur (1973) The Politics of World Hunger. New York: Harpers Magazine Press.

Singer, Hans W. "Distribution of Gains Between Investing and Borrowing Countries." American Economic Review, Vol. 40, May 1950, 472-92.

_____ (1964) International Development: Growth and Change. New York: McGraw-Hill Book Co.

_____ (1977) Technologies for Basic Needs. Geneva: International Labor Organization.

Singh, Harnam (1972) Studies in World Order. Delhi: Kitab Mahal (W.D.) Private Ltd.

Sinha, R., Peter Pearson, Gopul Kadekodi, and Mary Gregory (1978) Poverty, Income Distribution and Employment: A Case Study of India. Glasgow, Delhi, Oxford Project (Mimeo).

Sivard, Ruth Leger (1974) World Military and Social Expenditures. Leesburg, VA: WMSE Publications.

Slater, Courtenay "External Debt and Economic Development: Some Empirical Tests of Macroeconomic Approaches." Southern Economic Journal, Vol. 36, No. 3, January 1970, 252-62.

Slight, James and Burn, R. Scott (1858) The Book of Farm Implements and Machines. Edited by Henry Stephens, Edinburgh.

Smith, Miranda (1978) "The Greening of the South Bronx." Self-Reliance, January-February, pp. 13-15.

Smith, T.C. (1955) Political Change and Industrial Development in Japan: Government Enterprise, 1868-1880. Palo Alto, CA: Stanford University Press.

Smith, Vernon L. (1961) Investment and Production. Cambridge, MA: Harvard University Press.

Solo, R.A. (1975) Organizing Science for Technology Transfer in Economic Development. East Lansing, MI: Michigan State University Press.

Solow, Robert M. (1963) Capital Theory and the Rate of Return. Amsterdam: North Holland Publishing.

_____ "A Contribution to the Theory of Economic Growth." Quarterly Journal of Economics, 1956.

_____ (1967) "Interest Rate and Transition Between Techniques." Socialism, Capitalism and Economic Growth: Essays Presented to Maurice Dobb. Edited by H.G. Feinstein. Cambridge: Cambridge University Press.

_____ (1960) "Investment and Technical Change." In Mathematical Methods in the Social Sciences, edited by K.J. Arrow, S. Karlin and P. Suppes. Palo Alto, CA: Stanford University Press.

_____ "Technical Change and the Aggregate Production Function." Review of Economics and Statistics, 39, June 1957, 312-20.

Southworth, Herman M. and Johnston, Bruce F., eds. (1967) Agricultural Development and Economic Growth. Ithaca, NY: Cornell University Press.

Spence, Clark C. God Speed the Plw, The Coming of Steam Cultivation to Great Britain. Urbana, IL: University of Illinois Press.

Spengler, Joseph J. "Economic Factors in Economic Development." American Economic Review, Vol. 47, May 1957, 42-56.

Spenser, D.L. (1970) Technology Gap in Perspective. New York: Spartan Books.

Spregelgles, Stephen and Welsh, Charles, eds. (1970) Economic Development and Promise. Englewood Cliffs, NJ: Prentice-Hall, Inc.

Sraffa, P. (1960) Production of Commodities by Means of Commodities. Cambridge: Cambridge University Press.

Staley, Eugene and Morse, Richard (1965) Small Industry for Developing Countries. New York: McGraw-Hill Book Co.

Stanford Research Institute (1977) Solar Energy in America's Future: A Preliminary Assessment. Washington, D.C.: U.S. Government Printing Office.

Stanley, Dick (1978) The Arusha Windmill: A Construction Manual. Mt. Rainier, MD: VITA.

_____ Technology and Human Values. Santa Barbara, CA: Center for the Study of Democratic Institutions.

Stavrianos, L.S. (1976) The Promise of the Coming Dark Age. San Francisco, CA: W.H. Freeman.

Stein, B. (1974) Size, Efficiency and Community Enterprise. Cambridge, MA: Center for Community Economics Development.

Stern, A., and Stern, Robert L., ed. (1967) Technology and World Trade. Washington, D.C. U.S. Government Printing Office.

Stewart, Frances (1977a) "International Mechanisms for the Promotion of Appropriate Technology." Expert Meeting on International Action for Appropriate Technology.

_____ (1977b) Technology and Underdevelopment. Boulder, CO: Westview Press.

Stiglitz, J. (1972) Alternative Theories of Wage Determination and Unemployment in LDCs: The Labor Turnover Mode. Palo Alto, CA: Stanford University Press.

_____ "Distribution of Wealth and Income Among Individuals." Econometrica, 3, July 1969, 382-439.

_____ (1974) "The Efficiency Wage Hypothesis: Surplus Labor and the Distribution of Income in LDCs." Stanford University Economics Series, Technical Report, No. 152. Palo Alto, CA: Stanford University Press.

_____ and Shell, K. "The Allocation of Investment in a Dynamic Economy." Quarterly Journal of Economics, Vol. LXXXI, November 1967, 592-609.

Stokes, Bruce (1978) Local Responses to Global Problems: A Key to Meeting Basic Human Needs. Washington, D.C.: World Watch Institute, 1978, Worldwatch Paper # 17.

Stolper, Wolfgang F. (1966) Planning Without Facts: Lessons in Resource Allocation from Nigeria's Development. Cambridge, MA: Harvard University Press.

Strassman, W.P. "Interrelated Industries and the Rate of Technological Change." Review of Economic Studies, Vol. XXVII, October 1959.

_____ (1968) Technological Change and Economic Development: The Manufacturing Experience of Mexico and Puerto Rico. Ithaca, NY: Cornell University Press.

_____ and McConnaughey, John S. (1972) Appropriate Technology for Residential Construction in Less Developed Countries: A Survey of Research Trends and Possibilities. East Lansing, MI: Michigan State University Press.

Strathclyde University (1975) A Report of a Pilot Investigation of the Choice of Technology in Developing Countries.

Subrahmanian, K.K. (1972) Imports of Capital and Technology: A Study of Foreign Collaborations in India Industry. People's Publishing House.

Sundquist, James L. (1975) Dispersing Population: What America Can Learn from Europe. Washington, D.C.: The Brookings Institution.

Swamy, Dalip S. "Statistical Evidence on Balanced and Unbalanced Growth." Review of Economics and Statistics, Vol. 49, August 1967, 288-303.

Tangri, Shanti and Gray, H. Peter, eds. (1967) Capital Accumulation and Economic Development. Lexington: D.C. Heath and Co.

Taussig, F.W. (1915) Some Aspects of the Tariff Question. Cambridge, MA: Harvard University Press.

Taylor, John "On the Comparative Merits of Different Modes of Reaping Grain." Transactions of the Highland and Agricultural Society of Scotland, July 1844.

Taylor, Richard K. (1973) Economics and the Gospel. Philadelphia: United Church Press.

_____ Technology and Human Values. Santa Barbara, CA: Center for the Study of Democratic Institutions.

Temin, Peter (1964) Iron and Steel in Nineteenth Century America. Cambridge, MA: MIT Press.

Theberge, James D., ed. (1968) Economics of Trade and Development. New York: Johnson & Sons, Inc.

Theoblad, Robert "Technology in Focus: The Amerging Nations: Long-Term Prospects and Problems." Technology and Culture. III. Fall 1962, 601-16.

Thompson, .F.M.L. (1963) English Landed Society in the Nineteenth Century. London: Routledge and Kegan Paul.

Tilton, J.E. (1971) International Diffusion of Technology: The Case of Semiconductors. Washington, D.C.: The Brookings Institution.

Timmer, C. Peter, et al. (1975) The Choice of Technology in Developing Countries: Some Cautionary Tales. Cambridge, MA: Harvard University Press.

Tinbergen, Jan (1954) Centralization and Decentralization in Economic Policy. Amsterdam: North-Holland Publishing Co.

_____ (1958a) "Choice of Technology in Industrial Planning." Industrialization and Productivity, Bulletin No. 1, April 1958, 24-34.

_____ (1958b) The Design of Development. Baltimore, MD: Johns Hopkins University Press.

_____ (1975) Income Distribution: Analysis and Policies. Amsterdam: North-Holland Publishing Co.

_____ (1976) Reshaping the International Order. New York: E.P. Dutton.

_____ "Social Factors in Economic Development." Zeitscheift fur die Gesamte Staatswissenschaft, Vol. 124, September 1968.

_____ andBos, H.C. (1962) Mathematical Models of Economic Growth. New York: McGraw-Hill.

Tobin, James "Economic Growth as an Objective of Government Policy." American Economic Review, Vol. 54, May 1964, 1-20.

Todaro, M.P. "A Technological Note on Labor as an "Inferior" Factor in

Less Developed Countries." Journal of Development Studies, Vol. 5, July 1969.

Todd, Nancy, ed. (1977) The Book of the New Alchemists. New York: E.P. Dutton.

Toynbee, Arnold J. (1948) Civilization on Trial. New York: Meridian Books.

Turnham, David (1971) The Employment Problem in Less Developed Countries. Paris: Organization for Economic Cooperation and Development.

United Nations Transfer and Adaption of Technology for Industrial Development. General Review Prepared by the Secretariat. New York: United Nations, April 18, 1963.

_____ "Use of Industrial Equipment in Underdeveloped Countries." Industrialization and Productivity. Bulletin, No. 4, April 1961.

United Nations Conference on Trade and Development (UNCTAD) (1972) Guidelines for the Study of the Transfer of Technology to Developing Countries. New York: United Nations.

_____ (1974) Major Issues Arising from the Transfer of Technology to Developing Countries. New York: United Nations.

_____ (1975) The Role of the Patent System in the Transfer of Technology to Developing Countries. Report Prepared Jointly by the United Nations Department of Economic and Social Affairs, the UNCTAD Secretariat, and the International Bureau of the World Intellectual Property Organization (TD/B/AC.11/10/Rev.1.). New York: United Nations Conference on Trade and Development.

_____ "The Transfer of Technology." Journal of World Trade Law, 4, September-October 1970, 692-718.

United Nations Industrial Development Organization (1972) Changing Attitudes and Perspectives in Developing Countries Regarding Technology Licensing (ID/WG, 130/3). Vienna: United Nations Industrial Development Organization.

_____ (1974) Industrial Equipment from Developing Countries. Vienna: United Nations Industrial Development Organization.

_____ (1974) Recycling Technologies. Vienna: United Nations Industrial Development Organization.

_____ (1975) Comparable Equipment and Technologies from Developing Countries. Vienna: United Nations Industrial Development Organization.

_____ (1972) Specification and Remuneration of Foreign Know-How (ID/WG.130/1). Vienna: United Nations Industrial Development Organization.

United Nations Secretary General (1963) Science and Technology for Development. 8 Vols. New York: United Nations.

Urguidi, Victor L. "Latin American Development, Foreign Capital, and the Transmittal of Technology." El Trimestre Economioc, XXIX, January-March 1962, 19-29.

United States Agency for International Development (1977) Proposal for a Program in Appropriate Technology. Washington, D.C.: G.P.O.

United States Congress, House Committee on International Relations (1977) Proposal for a Program in Appropriate Technology. Washington, D.C.: G.P.O. (Transmitted by the Agency for International Development Pursuant to Section 107 of the Foreign Assistance Act.)

United States Congress, Office of Technology Assessment (1977) Application of Solar Technology to Today's Energy Needs. Washington, D.C.: G.P.O.

United States Congress, Senate Select Committee on Small Business (1975) The Role of Small Business in Solar Energy, Research Development, and Demonstration. Washington, D.C.: G.P.O.

United States Congress, Senate Select Committee on Small Business and Committee on Interior and Insular Affairs (1977) Alternative Long-Range Energy Strategies, 2 Vols. Washington, D.C.: G.P.O.

United States Department of State (1976) Development Issues: U.S. Actions Affecting the Development of Low-Income Countries.

_____ (1976) The United States and the Third World.

United States Executive, The Domestic Council (1976) 1976 Report on National Growth and Development: The Changing Issues for National Growth. Washington, D.C.: G.P.O.

United States National Academy of Sciences – National Research Council (1976) Energy for Rural Development: Renewable Resources and Alternative Technologies for Developing Countries. Washington, D.C.: G.P.O.

_____ (1976) The Role of U.S. Engineering Schools in Development Assistance. Washington, D.C.: G.P.O.

_____ (1978) U.S. Science and Technology for Development: A Contribution to the 1979 U.N. Conference. Washington, D.C.: G.P.O.

_____ Board of Science and Technology for International Development (1977) Appropriate Technologies for Developing Countries. Washington, D.C.: G.P.O. (Report # 24, Richard S. Eckaus.)

United States National Center for Appropriate Technology (1976) Proposal for the National Center for Appropriate Technology, U.S. Community Services Administration. Washington, D.C.

United States National Science Foundation (1977c) Appropriate Technology – A Directory of Activities and Projects. Washington, D.C.: G.P.O. 1977. (Ann Becker, Cecil Cook, Jr., and Eugene Eccle.)

_____ (1977b) Appropriate Technology and Agriculture in the United States. Washington, D.C. (Ann Becker.)

_____ (1977a) Appropriate Technology in the United States – An Exploratory Study. Washington, D.C.: G.P.O.

Usui, Mikoto (1975) "Oligopoly, R. and D. and Licensing: A Reflection Towards a Fair Deal in Technology Transfer." Organization for Economic Cooperation and Development, Development Center, Occasional Paper No. 7. Paris: Organization for Economic Cooperation and Development.

Vaitsos, Constantine "Patents Revisited: Their Function in Developing Countries." The Journal of Development Studies, 9, October 1972, 71-98.

Vakil, C.N. (1963) Poverty and Planning. Bombay: Allied Publishers.

Valaskakis, Kimon, et al. (1978) The Conserver Society. New York: Harper & Row.

Vietorisz, Thomas (1962) "Processing Techniques for Industrial Development." Science, Technology, and Development, 12 Vols. Washington, D.C.: G.P.O., IV103-18.

Villard, Henry H. (1963) Economic Development, Rev. Ed. New York: Holt, Rinehart & Winston, Inc.

Wade, Nicholas "Green Revolution: Creators Still Quite Hopeful on World Food." Science, 185, September 6, 1974, 844-5.

_____ "New Alchemy Institute: Search for an Alternative Agriculture." Science, 187, February 28, 1975, 727-9.

Wakefield, Rowan A. and Stafford, Patricia (1977) "Appropriate Technology: What It Is and Where It Is Going." The Futurist, April, pp. 72-6.

Walker, Charles R., ed. (1962) Modern Technology and Civilization: An Introduction to Human Problems in the Machine Age. New York: McGraw-Hill.

Walters, A.A. (1963) "Cost and Production Function: An Econometric Survey." Econometrica, 1063.

Ward, Barbara (1973) A New Creation: Reflections on the Environmental Issue. Vatican City: Pontifical Commission on Justice and Peace.

_____ (1968) "Technological Change and World Market." Applied Science and World Economy. Paper presented to the Committee on Science and Astronautics, U.S. House of Representatives, for the ninth meeting of the Panel on Science and Technology, February 1968. Washington, D.C.: G.P.O., 7-19.

_____ andDubos, Rene (1976) Only One Earth: The Core and Maintenance of a Small Planet. New York: W.W. Norton & Co., Inc.

Ward, Richard J. "Absorbing More Labor in LDC Agriculture." Economic Development and Cultural Change, Vol. 17, January 1960, 178-88.

_____ (1967) The Challenge of Development. Chicago: Aldine Publishing Company.

Waterston, Albert (1968) Development Planning: Lessons of Experience. Baltimore, MD: Johns Hopkins Press.

Weiner, Myron, ed. (1966) Modernization: The Dynamics of Growth. New York: Basic Books.

Weisman, Steve, ed. (1970) The Trojan Horse. San Francisco, CA: Ramparts Press.

Weisskopf, Walter A. (1973) Alienation and Economics. New York: E.F. Dutton and Co.

Wells, H.G. (1961) The Outline of History. New York: Doubleday and Co., Garden City Books.

Wharton, Clifton R. "The Green Revolution, Cornucopia or Pandora's Box." Foreign Affairs, April 1969, 464-76.

White, Lawrence J. (1974) Technology, Employment, and Development: Selected Papers Presented to Two Conferences Sponsored by the Council for Asian Manpower Studies. Quezon City, Philippines: Council for Asian Manpower Studies.

Wilcox, Chair, et al. (1966) Economics of the World Trade, 2nd Ed. New York: Harcourt, Brace and World, Inc.

Williams, B.R., ed. (1973b) Science and Technology in Economic Growth. New York: John Wiley & Sons.

_____ (1973a) "Science and Technology in Economic Growth."

Proceedings of a Conference Held by the International Economic Association at St. Anton, Austria.

_____ (1967) Technology, Investment and Growth. London: Chapman and Hall.

Williams, Hal, ed. (1978) The Uses of Smallness. Emmaus, PA: Rodale Press.

Wilson, J.S.G. and Scheffer, C.T., eds. (1974) Multinational Enterprises. Leiden, The Netherlands: A.W. Sijthoff.

Winner, Langdon (1977) Autonomous Technology: Technics-out-of-Control as a Theme in Political Thought. Cambridge, MA: MIT Press.

Wittfogel, Karl A. (1957) Oriental Despotism. New Haven, CT: Yale University Press.

Wolf, Charles F. "Institutions and Economic Development." American Economic Review, Vol. 45, December 1955, 687-883.

_____ andSufrin, Sidney C. (1965) "Technological Change and a Technological Alternative." Capital Formation and Foreign Investment in Underdeveloped Areas. Syracuse, NY: Syracuse University Press, 37-43.

World Bank, Central Projects Staff (1978) Appropriate Technology and World Bank Assistance to the Poor. Washington, D.C.: World Bank.

Yadelman, Montague "Integrated Rural Development Projects: The Bank's Experience." Finance & Development, March 1977, 15-19.

Yaksick, Rudy (1978) The Relationship Between Appropriate Technology and Business Development: Some Issues and Questions. Cambridge, MA: Center for Community Economic Development.

Young, A. "Increasing Returns and Economic Progress." Economic Journal, Vol. XXVIII, March 1928.

Zimmerman, L.J. (1965) Poor Lands, Rich Lands, New York: Random House.

Name Index

247

Subject Index

About the Authors

ROMESH KUMAR DIWAN, Professor of Economics at Rensselaer Polytechnic Institute, earned his doctoral degree in economics from the University of Birmingham, United Kingdom. Prior to joining the faculty at Rensselaer, he taught at Panjab University (India), the University of Glasgow (Scotland), and the University of Hawaii. He has served as Consultant to the United Nations (UNCTAD) and held visiting positions with Washington University (St. Louis) and London School of Economics. Professor Diwan's articles and book reviews have appeared in numerous publications.

DENNIS LIVINGSTON is a political scientist at Marlboro College, Vermont. His articles on energy and environmental policy, marine affairs, appropriate technology, and alternative futures have appeared in Environment, Bulletin of the Atomic Scientists, American Journal of International Law, Public Policy, and Ekistics. He reviews science fiction for Futures, and is currently involved in research on political and technological decentralization.

Pergamon Policy Studies